Progress or Catastrophe

PROGRESS OR CATASTROPHE

The Nature of Biological Science and Its Impact on Human Society

Bentley Glass

CONVERGENCE

A Series Founded, Planned, and Edited by
Ruth Nanda Anshen

PRAEGER

PRAEGER SPECIAL STUDIES • PRAEGER SCIENTIFIC

New York • Philadelphia • Eastbourne, UK
Toronto • Hong Kong • Tokyo • Sydney

Library of Congress Cataloging in Publication Data

Glass, Bentley, 1906–
Progress or catastrophe.

(Convergence)
Includes index.
1. Biology—Social aspects. I. Title.
II. Series: Convergence (New York, N.Y.)
QH333.G57 1985 304.2′7 85-3645
ISBN 0-03-001747-5
ISBN 0-03-001748-3 (pbk.)

Published in 1985 by Praeger Publishers
CBS Educational and Professional Publishing, a Division of CBS Inc.
521 Fifth Avenue, New York, NY 10175 U.S.A.

56789 052 987654321

Printed in the United States of America on acid-free paper.

INTERNATIONAL OFFICES

Orders from outside the United States should be sent to the appropriate address listed below. Orders from areas not listed below should be placed through CBS International Publishing, 383 Madison Ave., New York, NY 10175 USA

Australia, New Zealand
Holt Saunders, Pty, Ltd., 9 Waltham St., Artarmon, N.S.W. 2064, Sidney, Australia

Canada
Holt, Rinehart & Winston of Canada, 55 Horner Ave., Toronto, Ontario, Canada M8Z 4X6

Europe, the Middle East, & Africa
Holt Saunders, Ltd., 1 St. Anne's Road, Eastbourne, East Sussex, England BN21 3UN

Japan
Holt Saunders, Ltd., Ichibancho Central Building, 22-1 Ichibancho, 3rd Floor, Chiyo-daku, Tokyo, Japan

Hong Kong, Southeast Asia
Holt-Saunders Asia, Ltd., 10 Fl, Intercontinental Plaza, 94 Granville Road, Tsim Sha Tsui East, Kowloon, Hong Kong

Manuscript submissions should be sent to the Editorial Director, Praeger Publishers, 521 Fifth Avenue, New York, NY 10175 USA

CONVERGENCE

A Series Founded, Planned, and Edited by Ruth Nanda Anshen

Books in the Convergence Series

The Double-Edged Helix
Genetic Engineering in the Real World
Liebe F. Cavalieri

Progress or Catastrophe
The Nature of Biological Science and Its Impact on Human Society
Bentley Glass

Emerging Cosmology
Bernard Lovell

Creation and Evolution
Myth or Reality?
Norman D. Newell

Anatomy of Reality
Merging of Intuition and Reason
Jonas Salk

Science and Moral Priority
Merging Mind, Brain, and Human Values
Roger Sperry

Contents

Progress or Catastrophe

Convergence

Ruth Nanda Anshen

"There is no use trying," said Alice; "one *can't* believe impossible things."

"I dare say you haven't had much practice," said the Queen, "When I was your age, I always did it for half an hour a day. Why, sometimes I've believed as many as six impossible things before breakfast."

This commitment is an inherent part of human nature and an aspect of our creativity. Each advance of science brings increased comprehension and appreciation of the nature, meaning, and wonder of the creative forces that move the cosmos and created man. Such openness and confidence lead to faith in the reality of possibility and eventually to the following truth: "The mystery of the universe is its comprehensibility."

When Einstein uttered that challenging statement, he could have been speaking about our relationship with the universe. The old division of the Earth and the Cosmos into objective processes in space and time and mind in which they are mirrored is no longer a suitable starting point for understanding the universe, science, or ourselves. Science now begins to focus on the convergence of man and nature, on the framework which makes us, as living beings, dependent parts of nature and simultaneously makes nature the object of our thoughts and actions. Scientists can no longer confront the universe as objective observers. Science recognizes the participation of man with the universe. Speaking quantitatively, the universe is largely indifferent to what happens in man. Speaking qualitatively, nothing happens in man that does not have a bearing on the elements which constitute the universe. This gives cosmic significance to the person.

Nevertheless, all facts are not born free and equal. There exists a hierarchy of facts in relation to a hierarchy of values. To arrange the facts rightly, to differentiate the important from the trivial, to see their bearing in relation to each other and to evaluational criteria, requires a judgment which is intuitive as well as empirical. We need meaning in addition to information. Accuracy is not the same as truth.

Our hope is to overcome the cultural *hubris* in which we have been living. The scientific method, the technique of analyzing, explaining, and classifying, has demonstrated its inherent limitations. They arise because, by its intervention, science presumes to alter and fashion the object of its investigation. In reality, method and object can no longer be separated. The outworn Cartesian, scientific world view has ceased to be scientific in the most profound sense of the word, for a common bond links us all—man, animal, plant, and galaxy—in the unitary principle of all reality. For the self without the universe is empty.

This universe of which we human beings are particles may be defined as a living, dynamic process of unfolding. It is a breathing universe, its respiration being only one of the many rhythms of its life. It is evolution itself. Although what we observe may seem to be a community of separate, independent units, in actuality these units are made up of subunits, each with a life of its own, and the subunits constitute smaller living entities. At no level in the hierarchy of nature is independence a reality. For that which lives and constitutes matter, whether organic or inorganic, is dependent on discrete entities that, gathered together, form aggregates of new units which interact in support of one another and become an unfolding event, in constant motion, with ever-increasing complexity and intricacy of their organization.

Are there goals in evolution? Or are there only discernible patterns? Certainly there is a law of evolution by which we can explain the emergence of forms capable of activities which are indeed novel. Examples may be said to be the origin of life, the emergence of individual consciousness, and the appearance of language.

The hope of the concerned authors in Convergence is that they will show that evolution and development are interchangeable and that the entire system of the interweaving of man, nature, and the universe constitutes a living totality. Man is searching for his legitimate place in this unity, this cosmic scheme of things. The meaning of this cosmic scheme—if indeed we can impose meaning on the mystery and majesty of nature—and the extent to which we can assume responsibility in it as uniquely intelligent beings, are supreme questions for which this Series seeks an answer.

Inevitably, toward the end of a historical period, when thought and custom have petrified into rigidity and when the elaborate machinery of civilization opposes and represses our more noble qualities, life stirs again beneath the hard surface. Nevertheless, this attempt to define the purpose of Convergence is set forth with profound trepidation. We are living in a period of extreme darkness. There is moral atrophy, destructive radiation within us, as we watch the collapse of values hitherto cherished—but now betrayed. We seem to be face to face with an apocalyptic destiny. The anomie, the chaos, surrounding us produces an almost lethal distintegration of the person, as well as ecological and demographic disaster. Our situation is desperate. And there is no glossing over the deep and unresolved tragedy that fills our lives. Science now begins to question its premises and tells us not only what *is*, but what *ought* to be; *pre*scribing in addition to *de*scribing the realities of life, reconciling order and hierarchy.

My introduction to Convergence is not to be construed as a prefatory essay to each individual volume. These few pages attempt to set forth the general aim and purpose of this Series. It is my hope that this statement will provide the reader with a new orientation in his thinking, one more specifically defined by these scholars who have been invited to participate in this intellectual, spiritual, and moral endeavor so desperately needed in our time. These scholars recognize the relevance of the nondiscursive experience of life which the discursive, analytical method alone is unable to convey.

The authors invited to Convergence Series acknowledge a structural kinship between subject and object, between living and nonliving matter, the immanence of the past energizing the present and thus bestowing a promise for the future. This kinship has long been sensed and experienced by mystics. Saint Francis of Assisi described with extraordinary beauty the truth that the more we know about nature, its unity with all life, the more we realize that we are one family, summoned to acknowledge the intimacy of our familial ties with the universe. At one time we were so anthropomorphic as to exclude as inferior such other aspects of our relatives as animals, plants, galaxies, or other species—even inorganic matter. This only exposed our provincialism. Then we believed there were borders beyond which we could not, must not, trespass. These frontiers have never existed. Now we are beginning to recognize, even take pride in, our neighbors in the Cosmos.

Human thought has been formed through centuries of man's consciousness, by perceptions and meanings that relate us to nature. The smallest living entity, be it a molecule or a particle, is at the same time present in the structure of the Earth and all its inhabitants, whether human or manifesting themselves in the multiplicity of other forms of life.

Today we are beginning to open ourselves to this evolved experience of consciousness. We keenly realize that man has intervened in the evolutionary process. The future is contingent, not completely prescribed, except for the immediate necessity to evaluate in order to live a life of integrity. The specific gravity of the burden of change has moved from genetic to cultural evolution. Genetic evolution itself has taken millions of years; cultural evolution is a child of no more than twenty or thirty thousand years. What will be the future of our evolutionary course? Will it be cyclical in the classical sense? Will it be linear in the modern sense? Yet we know that the laws of nature are not linear. Certainly, life is more than mere endless repetition. We must restore the importance of each moment, each deed. This is impossible if the future is nothing but a mechanical extrapolation of the past. Dignity becomes possible only with choice. The choice is ours.

In this light, evolution shows man arisen by a creative power inherent in the universe. The immense ancestral effort that has borne man invests him with a cosmic responsibility. Michelangelo's image of Adam created at God's command becomes a more intelligent symbol of man's position in the world than does a description of man as a chance aggregate of atoms or cells. Each successive stage of emergence is more comprehensive, more meaningful, more fulfilling, and more converging, than the last. Yet a higher faculty must always operate through the levels that are below it. The higher faculty must enlist the laws controlling the lower levels in the service of higher principles, and the lower level which enables the higher one to operate through it will always limit the scope of these operations, even menacing them with possible failure. All our higher endeavors must work through our lower forms and are necessarily exposed thereby to corruption. We may thus recognize the cosmic roots of tragedy and our fallible human condition. Language itself as the power of universals, is the basic expression of man's ability to transcend his environment and to transmute tragedy into a moral and spiritual triumph.

This relationship, this convergence, of the higher with the lower applies again when an upper level, such as consciousness or freedom, endeavors to reach beyond itself. If no higher level can be accounted for by the operation of a lower level, then no effort of ours can be truly creative in the sense of establishing a higher principle not intrinsic to our initial condition. And establishing such a principle is what all great art, great thought, and great action must aim at. This is indeed how these efforts have built up the heritage in which our lives continue to grow.

Has man's intelligence broken through the limits of his own powers? Yes and no. Inventive efforts can never fully account for their success, but the story of man's evolution testifies to a creative power that goes beyond that which we can account for in ourselves. This power can make us surpass ourselves. We exercise some of it in the simple act of acquiring knowledge and holding it to be true. For, in doing so, we strive for intellectual control over things outside ourselves, in spite of our manifest incapacity to justify this hope. The greatest efforts of the human mind amount to no more than this. All such acts impose an obligation to strive for the ostensibly impossible, representing man's search for the fulfillment of those ideals which, for the moment, seem to be beyond his reach. For the good of a moral act is inherent in the act itself and has the power to ennoble the person who performs it. Without this moral ingredient there is corruption.

The origins of one person can be envisaged by tracing that person's family tree all the way back to the primeval specks of protoplasm in which his first origins lie. The history of the family tree converges with everything that has contributed to the making of a human being. This segment of evolution is on a par with the history of a fertilized egg developing into a mature person, or the history of a plant growing from a seed; it includes everything that caused that person, or that plant, or that animal, or even that star in a galaxy, to come into existence. Natural selection plays no part in the evolution of a single human being. We do not include in the mechanism of growth the possible adversities which did not befall it and hence did not prevent it. The same principle of development holds for the evolution of a single human being; nothing is gained in understanding this evolution by considering the adverse chances which might have prevented it.

In our search for a reasonable cosmic view, we turn in the first place to common understanding. Science largely relies for its subject matter on a common knowledge of things. Concepts of life and death, plant and animal, health and sickness, youth and age, mind and body, machine and technical processes, and other innumerable and equally important things are commonly known. All these concepts apply to complex entities, whose reality is called into question by a theory of knowledge which claims that the entire universe should ultimately be represented in all its aspects by the physical laws governing the inanimate substrate of nature. "Technological inevitability" has alienated our relationship with nature, with work, with other human beings, with ourselves. Judgment, decision, and freedom of choice, in other words *knowledge* which contains a moral imperative, cannot be ordered in the form that some technological scientists believe. For there is no mechanical ordering, no exhaustive set of permutations or combinations that can perform the task. The power which man has achieved through technology has been transformed into spiritual and moral impotence. Without the insight into the nature of *being*, more important than *doing*, the soul of man is imperilled. And those self-transcendent ends that ultimately confer dignity, meaning and identity on man and his life constitute the only final values worth pursuing. The pollution of consciousness is the result of mere technological efficiency. In addition, the authors in this Series recognize that the computer in itself can process information—not meaning. Thus we see on the stage of life no moral actors, only anonymous events.

Our new theory of knowledge, as the authors in this Series try to demonstrate, rejects this claim and restores our respect for the immense range of common knowledge acquired by our experience of convergence. Starting from here, we sketch out our cosmic perspective by exploring the wider implications of the fact that all knowledge is acquired and possessed by relationship, coalescence, convergence.

We identify a person's physiognomy by depending on our awareness of features that we are unable to specify, and this amounts to a convergence in the features of a person for the purpose of comprehending their joint meaning. We are also able to read in the features and behavior of a person the presence of moods, the gleam of intelligence, the response to animals or a sunset or a fugue by Bach, the signs of sanity, human responsibility, and experience. At a lower level, we comprehend by a similar mechanism the body of a person and understand the functions of the physiological mechanism. We know that even physical theories constitute in this way the processes of inanimate nature. Such are the various levels of knowledge acquired and possessed by the experience of convergence.

The authors in this Series grasp the truth that these levels form a hierarchy of comprehensive entities. Inorganic matter is comprehended by physical laws; the mechanism of physiology is built on these laws and enlists them in its service. Then, the intelligent behavior of a person relies on the healthy functions of the body and, finally, moral responsibility relies on the faculties of intelligence directing moral acts.

We realize how the operations of machines, and of mechanisms in general, rely on the laws of physics but cannot be explained, or accounted for, by these laws. In a hierarchic sequence of comprehensive levels, each higher level is related to the levels below it in the same way as the operations of a machine are related to the particulars, obeying the laws of physics. We cannot explain the operations of an upper level in terms of the particulars on which its operations rely. Each higher level of integration represents, in this sense, a higher level of existence, not completely accountable by the levels below it yet including these lower levels implicitly.

In a hierarchic sequence of comprehensive levels each higher level is known to us by relying on our awareness of the particulars on the level below it. We are conscious of each level by internalizing its particulars and mentally performing the integration that constitutes it. This is how all experience, as well as all knowledge, is based on convergence, and this is how the consecutive stages of convergence form a continuous transition from the understanding of the inorganic, the inanimate, to the comprehension of man's moral responsibility and participation in the totality, the organismic whole, of all reality. The sciences of the subject-object relationship thus pass imperceptibly into the metascience of the convergence of the subject and object interrelationship, mutually altering each other. From the minimum of convergence, exercised in a physical observation, we move without a break to the maximum of convergence, which is a total commitment.

"The last of life, for which the first was made, is yet to come." Thus, Convergence has summoned the world's most concerned thinkers to rediscover the experience of *feeling*, as well as of thought. The convergence of all forms of reality presides over the possible fulfillment of self-awareness—not the isolated, alienated self, but rather the participation in the life process with other lives and other forms of life. Convergence is a cosmic force and may possess liberating powers allowing man to become what he is, capable of freedom, justice, love. Thus man experiences the meaning of grace.

A further aim of this Series is not, nor could it be, to disparage science. The authors themselves are adequate witness to this fact. Actually, in viewing the role of science, one arrives at a much more modest judgment of its function in our whole body of knowledge. Original knowledge was probably not acquired by us in the active sense; most of it must have been given to us in the same mysterious way we received our consciousness. As to content and usefulness, scientific knowledge is an infinitesimal fraction of natural knowledge. Nevertheless, it is knowledge whose structure is endowed with beauty because its abstractions satisfy our urge for specific knowledge much more fully than does natural knowledge, and we are justly proud of scientific knowledge because we can call it our own creation. It teaches us clear thinking, and the extent to which clear thinking helps us to order our sensations is a marvel which fills the mind with ever new and increasing admiration and awe. Science now begins to include the realm of human values, lest even the memory of what it means to be human be forgotten. In fact, it may well be that science has reached the limits of the knowable and may now be required to recognize its inability to penetrate into the caprice and the mystery of the soul of the atom.

Organization and energy are always with us, wherever we look, on all levels. At the level of the atom organization becomes indistinguishable from form, from order, from whatever the forces are that held the spinning groups of ultimate particles together in their apparent solidity. And now that we are at the atomic level, we find that modern physics has recognized that these ultimate particles are primarily electrical charges, and that mass is therefore a manifestation of energy. This has often been misinterpreted by idealists as meaning that matter has somehow been magicked away as if by a conjuror's wand. But nothing could be more untrue. It is impossible to transform matter into spirit just by making it thin. Bishop Berkeley's views admit of no refutation but carry no conviction nevertheless. However, something has happened to matter. It was only separated from form because it seemed too simple. Now we realize that, and this is a revolutionary change; we cannot separate them. We are now summoned to cease speaking of Form and Matter and begin to consider the convergence of Organization and Energy. For the largest molecule we know and the smallest living particles we know overlap. Such a cooperation, even though far down at the molecular level, cannot but remind us of the voluntary cooperation of individual human beings in maintaining patterns of society at levels of organization far higher. The tasks of Energy and Organization in the making of the universe and ourselves are far from ended.

No individual destiny can be separated from the destiny of the universe. Alfred North Whitehead has stated that every event, every step or process in the universe, involves both effects from past situations and the anticipation of future potentialities. Basic for this doctrine is the assumption that the course of the universe results from a multiple and never-ending complex of steps developing out of one another. Thus, in spite of all evidence to the contrary, we conclude that there is a continuing and permanent energy of that which is not only man but all life. For not an atom stirs in matter, organic and inorganic, that does not have its cunning duplicate in mind. And faith in the convergence of life with all its multiple manifestations creates its own verification.

We are concerned in this Series with the unitary structure of all nature. At the beginning, as we see in Hesiod's *Theogony* and in the Book of Genesis, there was a primal unity, a state of fusion in which, later, all elements become separated but then merge again. However, out of this unity there emerge, through separation, parts of opposite elements. These opposites intersect or reunite, in meteoric phenomena or in individual living things. Yet, in spite of the immense diversity of creation, a profound underlying convergence exists in all nature. And the principle of the conservation of energy simply signifies that there is a *something* that remains constant Whatever fresh notions of the world may be given us by future experiments, we are certain beforehand that something remains unchanged which we may call *energy*. We now do not say that the law of nature springs from the invariability of God, but with that curious mixture of arrogance and humility which scientists have learned to put in place of theological terminology, we say instead that the law of conservation is the physical expression of the elements by which nature makes itself understood by us.

The universe is our home. There is no other universe than the universe of all life including the mind of man, the merging of life with life. Our consciousness is evolving, the primordial principle of the unfolding of that which is implied or contained in all matter and spirit. We ask: Will the central mystery of the cosmos, as well as man's awareness of and participation in it, be unveiled, although forever receding, asymptotically? Shall we perhaps be able to see all things, great and small, glittering with new light and reborn meaning, ancient but now again relevant in an iconic image which is related to our own time and experience?

The cosmic significance of his panorama is revealed when we consider it as the stages of an evolution that has achieved the rise of man and his consciousness. This is the new plateau on which we now stand. It may seem obvious that the succession of changes, sustained through a thousand million years, which have transformed microscopic specks of protoplasm into the human race, has brought forth, in so doing, a higher and altogether novel kind of being capable of compassion, wonder, beauty and truth, although each form is as precious, as sacred, as the other. The interdependence of everything with everything else in the totality of being includes a participation of nature in history and demands a participation of the universe.

The future brings us nothing, gives us nothing; it is we who in order to build it have to give it everything, our very life. But to be able to give, one has to possess; and we possess no other life, no living sap, than the treasures stored up from the past and digested, assimilated, and created afresh by us. Like all human activities, the law of growth, of evolution, of convergence draws its vigor from a tradition which does not die.

At this point, however, we must remember that the law of growth, of evolution, has both a creative and a tragic nature. This we recognize as a degenerative process, as devolution. Whether it is the growth of a human soul or the growth of a living cell or of the universe, we are confronted not only with fulfillment but with sacrifice, with increase and decrease, with enrichment and diminution. Choice and decision are necessary for growth and each choice, each decision, excludes certain potentialities, certain potential realities. But since these unactualized realities are part of us, they possess a right and command of their own. They must avenge themselves for their exclusion from existence. They may perish and with them all the potential powers of their existence, their creativity. Or they may not perish but remain unquickened within us, repressed, lurking, ominous, swift to invade in some disguised form our life process, not as a dynamic, creative, converging power, but as a necrotic, pathological force. If the diminishing and the predatory processes co-mingle, atrophy and even death in every category of life ensue. But if we possess the maturity and the wisdom to accept the necessity of choice, of decision, or order and hierarchy, the inalienable right of freedom and autonomy, then, in spite of its tragedy, its exclusiveness, the law of growth endows us with greatness and a new moral dimension.

Convergence is committed to the search for the deeper meanings of science, philosophy, law, morality, history, technology, in fact all the disciplines in a trans-disciplinary frame of reference. This Series aims to expose the error in that form of science which creates an unreconcilable dichotomy between the observer and the participant, thereby destroying the uniqueness of each discipline by neutralizing it. For in the end we would know everything but *understand nothing*, not being motivated by concern for any question. This Series further aims to examine relentlessly the ultimate premises on which work in the respective fields of knowledge rests and to break through from these into the universal principles which are the very basis of all specialist information. More concretely, there are issues which wait to be examined in relation to, for example, the philosophical and moral meanings of the models of modern physics, the question of the purely physico-chemical processes versus the postulate of the irreducibility of life in biology. For there is a basic correlation of elements in nature, of which man is a part, which cannot be separated, which compose each other, which converge, and alter each other mutually.

Certain mysteries are now known to us: the mystery, in part, of the universe and the mystery of the mind have been in a sense revealed out of the heart of darkness. Mind and matter, mind and brain, have converged; space, time, and motion are reconciled; man, consciousness, and the universe are reunited since the atom in a star is the same as the atom in man. We are homeward bound because we have accepted our convergence with the Cosmos. We have reconciled observer and participant. For at last we know that time and space are modes by which we think, but not conditions in which we live and have our being. Religion and science meld; reason and feeling merge in mutual respect for each other, nourishing each other, deepening, quickening, and enriching our experiences of the life process. We have heeded the haunting voice in the Whirlwind.

The Möbius Strip

The symbol found on the cover of each volume in Convergence is the visual image of *convergence*—the subject of this Series. It is a mathematical mystery deriving its name from Augustus Möbius, a German mathematician who lived from 1790 to 1868. The topological problem still remains unsolved mathematically.

The Möbius Strip has only one continuous surface, in contrast to a cylindrical strip, which has two surfaces—the inside and the ouside. An examination will reveal that the Strip, having one continuous edge, produces *one* ring, twice the circumference of the original Strip with one half of a twist in it, which eventually *converges with itself.*

Since the middle of the last century, mathematicians have increasingly refused to accept a "solution" to a mathematical problem as "obviously true," for the "solution" often then becomes the problem. For example, it is certainly obvious that every piece of paper has two sides in the sense that an insect crawling on one side could not reach the other side without passing around an edge or boring a hole through the paper. Obvious—but false!

The Möbius Strip, in fact, presents only one monodimensional, continuous ring having no inside, no outside, no beginning, no end. Converging with itself it symbolizes the structural kinship, the intimate relationship between subject and object, matter and energy, demonstrating the error of any attempt to bifurcate the observer and participant, the universe and man, into two or more systems of reality. All, all is unity.

I am indebted to Fay Zetlin, Artist-in-Residence at Old Dominion University in Virginia, who sensed the principle of convergence, of emergent transcendence, in the analogue of the Möbius Strip. This symbol may be said to crystallize my own continuing and expanding explorations into the unitary structure of all reality. Fay Zetlin's drawing of the Möbius Strip constitutes the visual image of this effort to emphasize the experience of coalescence.

R.N.A.

Introduction

This volume has grown out of a deepening concern, over more than fifty years as a scientist and teacher, for a wider understanding on the part of both my fellow biologists and the general public of the relation between scientific knowledge and discovery, on the one hand, and the course of social progress or the threat of catastrophe, on the other.

My early studies in genetics were chiefly focussed on the production of mutations by X-rays, and made me, like my teacher Hermann Joseph Muller, keenly aware of the hazards of high-energy radiation applied, even in small individual doses, to large numbers of people. In the 1930s and 1940s we thought principally of X-rays and of beta and gamma rays from radium used for medical diagnosis or treatment. Physicians—even most radiologists—at that time seemed to ignore or brush aside any concern about the damage to future children that might be done by such radiation, or even the danger of cancers appearing years after the exposure to radiation. In the 1950s there was a great public concern over possible damaging effects from nuclear fallout produced by the testing of nuclear weapons in the open atmosphere. A scrupulous effort to weigh these respective hazards led geneticists and radiologists to the conclusion that the risk of ill effects from medical radiation resulting in detrimental mutations and malignancies was many times greater than the risk of similar consequences resulting from fallout. I had a part in those debates and determinations of risk and genetic damage. Public concern led quickly to the treaty banning the atmospheric testing of nuclear weapons and to a reduction in the amounts of high-energy radiation used for medical diagnosis.

The Pugwash Conferences on Science and World Affairs between scientists from the East and the West represented another effort on the part of the scientific community itself to grapple in earnest with the perils of nuclear weapons that threaten all life on earth. Initiated in 1957 by an open appeal addressed to scientists by Albert Einstein and Bertrand Russell, these conferences did much to secure the nuclear weapons test ban and the nonproliferation treaty. I attended these conferences regularly for more than a decade and saw the agendas of the conferences gradually extend to include many problems of science and society other than the threat of nuclear warfare, although that always remained the primary concern of the gathered scientists. Other issues probed were the threats of chemical and biological warfare, the proper scope of science education and development especially in the Third World nations, the increasing pressures of the soaring world population, and the threats of insufficient food and of the exhaustion of irreplaceable natural resources.

In the late 1960s, at the State University of New York at Stony Brook, I organized a course in the Center for Continuing Education that reflected these and similar experiences. With the assistance of a number of scientists and humanists of the faculty, the course "Science and the Future of Man" provided lectures and group discussions on such topics as: the nature of science; the growth of science, its applications and their social consequences; the physical and biological destructiveness of nuclear weapons; the roles of medical research and the science of nutrition in reducing average death rates around the world and in insuring better health for all; the exponential increase of the world's population as a consequence, and the pressure on the food supply and living space itself; the "green revolution" and the prospects of increasing staple food supplies; the nature of ecosystems and their disturbance by human activities; the psychological consequences of living in an increasingly competitive world with an ever-mounting tempo of change; the possible conquest of mental diseases; the prospects for voluntarily regulated birth rates, external fertilization and implantation of selected embryos, cloning, genetic engineering, and the intimate interplay of heredity and environment in shaping every human characteristic all of which result from the development of human genetics and control over reproductive processes; and finally, the grave ethical issues and quandaries that face humanity at the dawn of the twenty-first century.

These problems all grow in large part from scientific developments and their technological applications, combined with human ignorance, desire, greed, and selfishness. For a decade, thousands of students in this course expressed to us teachers their intense concern over the human condition, their despair at the blindness of governments, and their hopes for a better future, when people everywhere would become more enlightened regarding the role of science in their lives and would assume a personal responsibility in charting the course of the future, in helping to make the right choices to prevent war or famine, ill health and mental disorientation, and the destruction of the environment in the world of their descendants. It is out of all these experiences that the present book has grown.

Each of these essays, with one exception, has been previously published in some scientific or general periodical. They have been brought up to date, but they were addressed to a variety of audiences at the time, and it has seemed best to me not to alter too much that sense of immediacy. At the same time, that circumstance accounts for some lack of a unified relation of each essay to the others and for some redundancy, which it is to be hoped the reader will forgive for the sake of emphasis upon crucial issues.

After a preliminary essay that provides a general perspective (see Part I), there follows a series of essays (see Part II) that deals with the nature of biological science: How do biologists undertake to explain nature? How do they see the relation of causes to effects? What is the relation of biology to the physical sciences? To what degree is evolution the central theme of modern biology? How can the study of human genetics and evolution lay a basis for greater racial understanding and tolerance? My own personal interest in the history of science accounts for an essay on the origin of the concept of *species*. For "species," in the biological sense, had to be defined before there could be any theory of organic evolution or any theory of the origin of species by means of natural selection.

In Part III, the first essays deal with the question of whether scientific discoveries and consequently social progress can continue indefinitely, and the essays also address the appropriate nature of education, particularly science education, during a time of rapid technological and social change. Other chapters deal with some of the major crises growing out of scientific technology: the overwhelming human multitude; the special concerns in science and education of the developing countries; the biology of nuclear war. The subjects discussed by no means cover all the important relations of science to society in our time. They reflect the limited knowledge of a particular biologist about these matters, knowledge all too inadequate after even a lifetime of study and reflection.

My own boyhood was spent in China, where I was born. Very early I became acquainted with the concept of the "yin" and the "yang" and with the symbol of their intimate, interlocked relationship. Yin/yang are commonly supposed, especially by Westerners, to represent simply the female and the male essences of nature. Actually, they signify much more: the relation of light to darkness; of good to evil; the complementary nature of opposites. To me, as a geneticist, the yin/yang symbol came to represent the intimate, indissociable interplay of heredity and environment in the life of a plant, an animal, a human being. In respect to the effects of science and technology, yin and yang may fittingly represent the Promethean and Pandoran consequences of any scientific advance. All our knowledge converges here.

In one respect, the symbol remains insufficient. The philosopher Hegel came closer to the truth when he enunciated his principle of "thesis-antithesis-synthesis." Our concern grows out of the consequences of what Niels Bohr called "complementarity." We make a grave mistake whenever we neglect either the yin or the yang, the thesis or the antithesis, the heredity or the environment. The best genes will do a child little good in certain environments. The best environment cannot make a human out of a monkey. We need equally to give attention to both sides of this complementarity of heredity by environment, of environment by heredity. For each human genotype there is indeed some optimum environment, though the elements of that optimum environment differ from person to person, from genotype to genotype. The complementation leads to a true synthesis, a new whole. We are even now only at the beginning of our great scientific endeavor in biology, to define justly the best conditions for development and for life in relation to the variable nature of our gene pools—to find in each case the fitting synthesis.

I lay myself open to the criticism that I suggest few remedies for those ills of science and society that escape from Pandora's Box. Yet for me to do so would be a false assumption of knowledge. To cry an alarm is the first need, to add another voice to those of others who have spoken out. To find and utilize socially acceptable and also effective remedies will be the continuing, joint task of natural science, social science, and humanism for centuries to come.

Part I
The Dilemma of Science and Technology

1 Prometheus and Pandora

In the myths of the Greeks, Prometheus, pitying the sad estate of mortals and defying the will of Zeus, stole fire from the sun and brought it to man. Out of this single gift there blossomed the early technologies of man as toolmaker, traveler, and food-grower. Hear the words of Aeschylus, in the play *Prometheus Bound*:

> I gave to mortals gifts.
> I hunted out the secret source of fire.
> I filled a reed therewith,
> fire, the teacher of all arts to men,
> the great way through....
> I made men cease to live with death in sight....
> Blind hopes I caused to dwell in them...
> To it I added the good gift of fire....
> Hear rather all that mortals suffered,
> Once they were fools. I gave them power to think.
> Through me they won their minds...
> Seeing they did not see, nor hearing hear.
> Like dreams they led a random life.
> They had no houses built to face the sun,
> of bricks or well-wrought wood,
> but like the tiny ant who has her home
> in sunless crannies deep down in the earth,

This chapter is based in part on an anniversary address to the New York Academy of Medicine in 1971 and in part on an address contributed to the 1967 International Edison Birthday Celebration Science Youth Day at Dearborn, Michigan, sponsored by the Thomas Alva Edison Foundation. It has been extensively revised, and is here reprinted with the permission of the two organizations named.

they lived in caverns.
The signs that speak of winter's coming,
of flower-faced spring, of summer's heat
with mellowing fruits,
were all unknown to them.
From me they learned the stars that tell the seasons,
their risings and their settings hard to mark.
And number, that most excellent device,
I taught to them, and letters joined in words.
I gave to them the mother of all arts,
hard working memory.
I too, first brought beneath the yoke
great beasts to serve the plow,
to toil in mortals' stead.
Up to the chariot I led the horse that loves the rein,
the glory of the rich man in his pride.
None else but I first found
The seaman's car, sail-winged, seadriven....
Listen, and you shall find more cause for wonder.
Best of all gifts I gave them was the gift of healing.
For if one fell into a malady
there was no drug to cure, no draught, or soothing ointment.
For want of these men wasted to a shadow
until I showed them how to use
the kindly herbs that keep us from disease....
Deep within the earth are hidden
precious things for men, brass and iron, gold and silver.
Would any say he brought these forth to light
until I showed the way?
No one except to make an idle boast.
All arts, all goods, have come to men through me.

In the Great Hall of the National Academy of Sciences in Washington there is a magnificent mural depicting the theft of the divine fire from the sun by Prometheus. He is reaching forward behind the hurtling chariot of Helios, the sungod, and stretches his hollow reed toward the fiery sun itself. Behind Prometheus stands Athena, goddess of wisdom, to aid him in his mission. Beneath the mural run those awesome words of the poet: "All arts, all goods, have come to men through me." It is thus a familiar thought, among scientists and technologists, that Prometheus symbolizes science and technology in the life of mankind.

For his defiance, and because the gods on Olympus feared that mankind, raised by knowledge to gain power and progress, would eventually displace the gods themselves, Prometheus was condemned to immortal torture, chained to a great rock and daily torn by the rage of Zeus's great eagle, only to regenerate his tissues during the darkness. That, and Prometheus's defiance, were Aeschylus's subject in his great poem. But the myth is not yet complete. The Greeks attached a second part, also well known to most of us, yet all too often severed from its essential connection to the tale of Prometheus.

It runs as follows. Woman did not exist until the gods, jealous of Prometheus's gifts to man, created her in the fullest beauty of Aphrodite. Hermes bestowed upon her the power of persuasion, and Apollo gave her music to entice the heart of man. The name of this first woman, sent by the gods to Prometheus and his brother Epimetheus and by them introduced to man, was Pandora. Among other characteristics with which Pandora had been endowed by the gods was the fatal gift of curiosity. In a jar or box in the house of Epimetheus the brothers had stored certain leftover articles for which, in fitting man out for a better life, they had had no need. Or, as another version of the myth has it, the box was one which the gods had given to Pandora and which contained, so they said, her dowry. Burning with curiosity to know the contents, Pandora opened the box in spite of stern warnings, and immediately there escaped a "multitude of plagues for hapless man—such as gout, rheumatism, and colic for his body, and envy, spite, and revenge for his mind." Too late, Pandora slammed shut the lid. Everything within had escaped save one last supposedly useless thing, which had been at the bottom. It was hope.

Like so many of the myths, these two are clearly inconsistent; but the Greeks never worried much about such things, since myths reflect only certain aspects of life, not the whole of it. In any case, it is the reconciliation, the unification, of these two myths in the life of modern man that I wish to emphasize.

Somehow the Greeks had grasped a great truth. For every gift of new knowledge that expands human power and enters the fabric of our civilization, there are offsetting plagues and worrisome problems. These are generated by the very gifts of the gods themselves.

Every new scientific discovery, every new advance in technology creates long-term effects and side-effects that detract substantially from whatever immediate benefits accrue. Until we can establish appropriate social institutions for technology assessment, to examine with great care and by scientific means what adverse consequences are likely to flow from technological innovation and how they may be circumvented, we will continue to be the children of Pandora. This is the real tragedy of Prometheus—not his defiance of the gods, nor yet his kindness of heart toward suffering humanity, but rather his blindness to the full spread of consequences.

Turning from myth to history, we see five hundred thousand years of inconceivably slow improvement in human tools succeeded by some fifty thousand years of more rapid, yet still almost equally imperceptible, progress and ending with a quite rapid change as the planting of seeds and the manufacture of metal made possible towns and wars alike. Important as was the sword, the pen, somewhat later, was even more important, for the keeping of records not only initiated recorded history, but also a far more effective transmission of the skill and lore of the past to future generations of men. Education began, and with schools each generation could build quickly on the totality of past achievement, could command the totality of past human power, and could explore previously hidden problems with new human curiosity instead of forever painfully rediscovering for themselves what their forebears had discovered in their own time.

The sources of modern power lie in modern science, upon which our engineering, technology, and medicine are based. One thousand years ago, there was a first glimmer of the new age when magnetism was discovered; yet not until the year 1600 did science begin to become an active pursuit of many people, working along systematic lines of experimentation and proof to discover the laws of nature. Our own time, in Carl Becker's words, is the era of "Instruments of Precision," the age when the exponential rate of increase of human power became reflected in an exponential increase of population and the total exploitation of earth by man.[1] As man now looks from his tiny planet out into the immensity of whirling galaxies, he is master of all on earth, and he dreams of flight across space and the settlement of other worlds; but he is also afraid. It is in truth of himself that he is afraid, for this is the age of human crisis, an age when human powers have become so vast that in an instant of anger or prejudice man can destroy himself and his entire planet.

CRISES OF OUR TIMES

The problems of our age are in great measure the risks and the quandaries arising from our newfound powers and our inability to find the honesty and courage, the insight and wisdom needed to deal with them safely. Most immediate among these is the crisis of war and peace, the ever-present threat of nuclear weapons that can destroy most of human civilization in a few moments, so thoroughly that it might never recover. Mankind has now lived with this threat for over thirty-five years, and young people have never known a world without it. Yet, as the possession of nuclear weapons by more and more nations becomes a reality, the threat is not lessening, but growing. What schemer or madman may engulf us in destruction? What new structure of society or politics can save us?

An equally imminent, though more protracted, danger is the explosion of human populations now taking place in the world. In spite of two worldwide wars and the occurrence of millions of horrible, needless deaths from war injuries, famine, and disease, the population of the world has doubled and redoubled in this century and will very likely do so again before the year 2000. Tremendous advances in the control of infant disease and mortality have not been followed, except in a few countries, by any great reduction in the birth rate. Many countries are still increasing in population at a rate of two or even three percent annually, doubling their populations in a third to a quarter of a century; and most countries exceed an annual

[1]Carl Becker, *Progress and Power* (Stanford, Cal.: Stanford University Press, 1936).

increase of one percent *per annum*, which is enough to double the population in seventy years.

The growth of population brings as a consequence other mortal perils, the exhaustion of resources, organic as well as mineral, and the pollution of the environment. Of these, the former is probably least critical. Energy from other sources may be ours in abundance before our coal and oil are gone. Yet the immediate gains from the development of nuclear energy for power must be balanced against the steady input to the terrestrial environment of tons of high-level radioactive wastes with physical half-lives of tens of thousands of years. Recourse to fast breeder reactors, that produce more fuel than they consume, will simply enhance the risks of human exposure to incredibly toxic plutonium, as well as that of hijacking by political extremists who want to make atomic bombs and threaten world order.

Unlimited power, with a minimum of any harmful byproducts or side effects, may come from controlled nuclear fusion, although probably not soon. Solar power, converted into hydrogen and oxygen as fuels by means of the electrolysis of water, may some day be a reality. But much enormously expensive research will have to be carried forward to achieve those goals, and only rich governments can afford the cost before there is any repayment. Solar cells distributed in satellites and beaming energy to the earth, or using it locally for manufacturing processes, may be the answer. That is Gerard O'Neill's favorite hope.[2]

Metals, wasted and dissipated beyond recovery, may be replaced by plastics. We will come to depend more on raw materials we can grow—on forests and crops—rather than upon materials we mine. We may ere long tap many of the almost inexhaustible stores of materials in the sea, provided energy is cheap enough. O'Neill is convinced that mining asteroids in space in order to secure minerals can be pursued most hopefully. Before that, however, the moon must be colonized and the "mass-driver" perfected, in order to build space colonies in appropriate locations.

[2]Gerard K. O'Neill, *2081: A Hopeful View of the Human Future* (New York: Simon & Schuster, 1981).

On the other hand, the pollution of our soil and the fresh-water supplies of our continents can go only so far before tragic results will follow. Already, the organic materials flowing into many streams and lakes have produced such a staggering growth of bacteria and algae that the dissolved oxygen of the waters has been exhausted, and most animal life has been eliminated. Chemical wastes have produced another kind of poisoning. Dioxin and PCB's (polychlorinated biphenyls), byproducts or organic chemical manufacturing processes, have poisoned soil adjacent to industrial dumps and have seeped into streams and floodways, to the great hazard of human life in such areas. Acid rain arising from the combustion of high-sulfur fuels has sterilized large bodies of inland waters and rendered them almost lifeless. Swimming and fishing and other forms of water recreation are spoiled. Heavy irrigation with saline waters, as in our Southwest, has impregnated the soil with salts that threaten to turn once bountiful fields into wasteland.

Air pollution threatens to be even more quickly disastrous. Smog still blankets most of the United States and is a serious threat to health in our greater cities. All this pollution, of course, increases in direct proportion to the increase in population, if not even more rapidly. For example, in America the number of cars helping to produce smog has in the twentieth century increased at a far faster rate than the population itself. The demand of modern societies for faster, abundant, cheap transportation creates a dilemma that can be solved only by a radical technological departure from our current devotion to the automobile.

Advanced agricultural practices add to our pollution of soil and water, and inadvertently create new problems of health and nutrition. Just a couple of examples will suffice. The trend toward the irrigation of arid lands by using either underground reservoirs of limited volume or by building great dams and reservoirs is epitomized by the history of the Aswan Dam and Reservoir in Egypt. The immediate gains in the production of food from new irrigated acres must be balanced against the spread through the irrigation channels of the snails that carry the dread disease of schistosomiasis—now Egypt's primary health problem—and the collapse of the Mediterranean fisheries that depended on the fertilizing effect of the Nile on the sea, as well as the greatly increased need for expensive artificial fertilizers for the lands deprived of the fertilizing Nile floods. The rapid increase in the population of Egypt fostered by the increased food supply has within less than two decades consumed every gain in production brought about by the new level of irrigation.

A second example followed the introduction of the new "Miracle Rice," which was produced by genetic selection in the Philippines and widely introduced in Southeast Asia. This highly productive rice, with its excellent resistance to falling flat and "lodging" in high winds, does require for its extra yield a high level of artificial fertilizer. The result has been to make the economies of the countries relying upon this rice more and more dependent upon the oil-producing Middle East, since they had little or no oil of their own. (Modern fertilizers are either produced from petroleum or require the energy of oil as fuel in their manufacture.)

Many long-term effects or side effects can be avoided by a combination of foresight and restraint. The foresight can be supplied by further development of science and its application to the technological problems. The restraint, on the other hand, must be governmental, since otherwise the selfish interests of those who benefit immediately from any new technological development will rule. But the governmental restraints are not likely to arise unless there is a new spirit among the people that demands such deliberate control.

The bounties of science, translated through technology into applications to the life of man, are not achieved without that supposedly fatal gift of curiosity, which is obviously no longer confined to woman, but is the predominating motivation of the scientist. It is not sufficient to possess memory, and through education to transmit the knowledge of the past to the future. Man grows in power through expanding his understanding of the nature of the world and of himself. Thus the pragmatic art of healing has been enlarged to become, in the past century and a half, the sciences of anatomy, physiology, pathology, and pharmacology, and to these more ancient studies have been added such altogether new insights as are provided by biochemistry, genetics, psychology, and ecology.

It is to the highest degree unfortunate that those who fund our institutions of science and learning have not yet comprehended that without a continuous search for new knowledge, without the repeated experimental testing of new and old observations and concepts, and without an unceasing synthesis of more accurate and reliable systems of thought, education becomes rote learning, and all progress ceases. It is appalling that modern legislators still think that education can be defined in terms of the number of "contact hours" spent by a teacher in the classroom, and that in years of tight budgets research should be the first item to be sacrificed. If the expansion of knowledge is not a fundamental part of education, if the teacher of a science is not himself to participate in the advancement of science, then we are indeed consigned to blind tradition, to capitalization upon the basis solely of past achievements. If, on the contrary, we believe that healing and the prevention of disease, better agriculture, and a balanced environment are important to human welfare, we must press forward to seek answers wherever our curiosity directs us and the needs of man are apparent.

PROGRESS AND PERIL IN MEDICINE AND PUBLIC HEALTH

In no aspect of human welfare is the role of scientific advance and developing technology more evident than in the areas of medicine and public health. Here, too, the peril of unheeded or unexpected side effects and the consequences of hurried efforts to introduce into practice and commercial profit what seem to be promising new discoveries mar the record of unselfish devotion to patient and populace. It is most instructive to look closely at these matters.

The past 125 years have witnessed the advent of anesthesia during surgery, of antiseptic and aseptic precautions, of antibiotics and other wonder drugs, of enormous improvements in diagnosis, and of immense advances in immunization. In 1851 the annual death rate in England and Wales was 22.0 per 1,000. In the United States today it is under 10 per 1,000. More indicative of the overall progress in public health are the comparative life expectancy figures: about 40 years at birth in 1850; well over 70 years in 1980. These advances were chiefly the result of applications of our expanding knowledge of sanitation, nutrition, immunization, and medical diagnosis and treatment, knowledge gained through research carried out in educational and scientific institutions.

The patterns of healing are changing radically in recent decades. The conquest of infectious diseases is well illustrated by the status of tuberculosis. Until 1900 tuberculosis was still the second-ranking cause of death, a disease of which my own mother was a victim. Improvements in public health and nutrition in the next half century cut the mortality rate from 100 per 100,000 to 25 per 100,000; but morbidity and economic loss remained high. In the 1950s new drugs were introduced to combat tuberculosis. First streptomycin, then isoniazid combined with streptomycin, produced a precipitous decline in mortality from tuberculosis, which no longer figures in the tables of the common causes of death. Most of the hospitals for tuberculosis have been closed. Tuberculosis, moreover, is but one of many infectious diseases which have been brought under control. Except for influenza, pneumonia, and other bronchopulmonary disorders, there are no bacterial or viral infections among the fifteen most common causes of death of Americans today.

True, the chapter is not closed. Mutations of microorganisms producing strains resistant to drugs and antibiotics or to immunizing sera may occur and create temporary flare-ups in the incidence of infectious diseases. Carelessness and improvidence contribute overwhelmingly to our current scandalous increase in venereal diseases and to a recrudescence of tuberculosis here and there. The high incidence of infant mortality in the United States, in comparison with some other Western countries, implies that our standards of prevention are not what they could be. The major health problems today, as reflected in our mortality figures, are the cardiovascular disorders, cancer, accidents, diabetes, cirrhosis of the liver, suicide, congenital malformations, and homicide, in about that relative ranking. Thus the causes of death are today more attributable to biochemical deficiencies, nutritional deficiencies and disorders, constitutional defects, psychological illness, and of course accidents and violence.

Human heredity plays a significant part in many or most of these. The deterioration of the environment too is leading to a rapidly rising but unmonitored contribution to illness and death from toxic materials in food, water, and air. Increasing exposure to high-energy radiations and to environmental carcinogens provokes an upward trend in the incidence of cancer. In short, the patterns of illness and death are quite different from those of even twenty-five years ago. Medical research must consequently pursue new directions. Medical education must be transformed to emphasize new issues and to supply new needs. Already, physicians have undergone a technological obsolescence strikingly confessed by the advocacy on the part of the American Medical Association of a policy that would ban permanent licensing and would require periodic retraining and reexamination of practitioners.

The very increase in the numbers and density of the human population in nearly all countries of the world is itself creating new health problems and is forcing us toward a reorientation of our thinking. What is the optimum population? How many people can the world support? Kingsley Davis, one of our foremost sociologists and demographers, pointed out in 1970 that an increase of the world's population at the present rate for only 200 more years would result in a total population of 135 billion people, or about 37 times the present number.[3] Quite aside from the ability of our food producers to nourish adequately any such number of people, there are other questions which must be answered. At such a density as would result, how can the volume of wastes produced by the population be disposed of? How will adequate supplies of pure water for drinking and washing be supplied? At what standard of living can such a population be maintained? The conclusion indeed seems inescapable; as Kingsley Davis says, "the earth's human multiplication has reached a climactic point, and. . . a change of trend, probably in this century, is inevitable."

All of us today are aware that the growth of the human population is not attributable to any increase in the birth rate, but to a remarkable decline in the death rate, a decline brought about by the improvements in public health and medicine. The conclusion which I want to stress from this point on is that, as

[3]Kingsley Davis, "The climax of population growth: Past and future perspective," *California Medicine* 113(1970):33–39.

in all matters of assessing technology, those who are responsible for any technological alteration of the environment that brings about certain consequences cannot morally limit their concern to immediate beneficial effects. The long-range effects, which may not be so desirable, are their responsibility, too. At least in our modern world, it is no longer possible to distinguish between Prometheus and Pandora!

Let me quote Kingsley Davis further:

> Increasingly, human survival is achieved by means that release rather than intensify genetic selection. The environment to which "adaptation" is being made is the artificial one created by ever more complex technology. The genius of this technology is that it makes biological adaptation unnecessary. Every demand made of the organism by the artificial environment is met, not by a genetic change suitable to satisfy that demand, but by a new alteration of the artificial environment. One might think that the achievement of an advanced civilization in providing rich nourishment, for example, would be met by evolutionary adaptation of the human organism to tolerate mass ingestions of sugars. On the contrary, diabetes is treated in such a way as to keep diabetics alive and reproducing, thus lowering rather than raising the pancreatic capacity of the population. At the very least, one might think that life in a sociocultural environment would place a premium on the ability to learn, thus evolving a more intelligent animal over time; but technology rises to this demand in another way, by using the production of the intelligent to support the unintelligent. As long as one in a million individuals has creative capacity, the process of improving the technological environment, thus releasing the evolutionary pressure on the organism, can continue.[4]

Kingsley Davis went on to say that therapeutic and preventive medicine were first added to economic development to bring this situation about, but in recent years medicine has to some extent actually substituted for economic development. Thus, in Mauritius, Ghana, Indonesia, and Morocco, to choose just a few examples, the 1950s showed a remarkable increase in

[4]Ibid., p. 34.

life expectancy, while at the same time the income per capita *declined* by a corresponding amount.

Even in America, it seems to me, the medical profession has not been openly and ardently committed to warning the public about the changing problems of health involved in the technological alteration of our environment. It has not been the community of physicians that has been foremost in analyzing and warning about the dangers of smog—the increased exposure of the population to carcinogens and to irritants that are already greatly increasing the frequency and severity of respiratory disorders in our population. Where is the medical pressure against the use of chlorinated hydrocarbons such as DDT? Of the increasing release into our fresh waters of toxic lead, mercury, and cadmium? Of the dangers to hearing produced by high noise levels and the psychological stresses of heavy urban crowding?

I recall with lingering dismay the impression made upon me, when I was a very young graduate student, by an address given by Hermann Joseph Muller to the McClennan County Medical Society in Waco, Texas, in which he pointed out the surely injurious and mutagenic effects stemming from exposures to X-rays, which he had shortly before, in 1927, demonstrated would induce lethal gene mutations at frequencies up to one thousand-fold above spontaneous rates. His words were met by ridicule and hostility from the users of X-rays for diagnostic purposes. Thirty years later, when both he and I were members of the National Academy of Sciences' Committee on the Genetic Effects of Atomic Radiations, evidence was found that the average exposure of Americans to diagnostic X-rays and radioactive elements was at least as great, at that date, as the spontaneous background exposures received during a lifetime, and vastly exceeded the exposure from radioactive fallout, which was so much a public concern at the same period. I am pleased to say that the evidence was accepted by noted radiologists of the country and that rigorous measures to cut down needless exposures were undertaken. The result has been most gratifying. There has been no loss in the use of diagnostic X-rays, but exposures have been greatly reduced through the development of faster film requiring shorter exposures, elimination of much fluoroscopy, better coning of X-ray beams and monitoring of machines, and better shielding of patients. Today the average exposure of Americans is estimated to be less than half what it was in 1956. That is truly a fine example of proper attention to the adverse effects of a certain scientific technology. Yet why did the medical profession wait thirty years to become concerned about the possible dangers they were themselves inflicting on the population? And why did that question have to be urged upon them by scientists outside their own profession?

As conditions requiring medical attention change, numerous questions arise in which there appears to be a conflict between an immediate advantage to some one and some disadvantage to society as a whole. The ethical questions that confront us will not be easy to resolve. Take death itself as an example. If, at some age or other, all men prove mortal, is it reasonable to try to overcome all causes of death? Would it not be better to select the quickest and most painless form of demise and leave that alone, while concentrating on the elimination of the most lingering, most debilitating, and most painful? Would this reasoning lead us to the conclusion that much of our research on cardiovascular disorders is unwise? Should we try merely to prevent premature death from cardiovascular disorders while concentrating on the control of cancer? Again, since there is a biologically predetermined species-specific life-span that in each species is determined by the action of natural selection on the genes, and that strikes a balance between the evolutionary values of a rapid reproductive turnover of genotypes available for selection and a prolonged lifetime, is it wise for us to try to prolong the average lifetime beyond an age when reproductive contributions may be made and the care and upbringing of the young are needed? But if the social utility and value of many elderly persons be argued, should we not concentrate our efforts on prolonging the years of physical and mental activity and useful contribution to society, rather than simply on extending the capacity for a vegetative existence? In a world increasingly overcrowded, in which the prolongation of four persons' lives by twenty years each has the same effect upon size of population as the birth of another baby, is the baby not more valuable to the population, even though it may, upon reaching adulthood, in turn contribute to an increase in the population?

Surgery is clearly on the eve of great developments in the replacement of diseased or defective parts. Studies in immunogenetics and radiation biology may well make it possible within the next decades to overcome the normal resistance to successful transplants of organs between persons who do *not* carry identical sets of genes. The discovery that human beings have a complex of histocompatibility genes like that of the mouse has led rapidly to laboratory typing of particular individual's histocompatibility genes. Like the older blood-typing procedures, these methods make it possible to determine which relatively unrelated donors of organs might safely be used for transplants to a person in need. There are vastly greater numbers of such persons like us in respect to our histocompatibility genes than our few close relatives, who in any case may not be suitable donors of organs or tissues because they do not share our particular histocompatibility genes. We may look forward to the time when, as a matter of anticipatory precaution, each of us will be typed while an infant for our histocompatibility genes as well as our blood groups. Then, in time of crisis, just as we now have blood banks with supplies of matching blood, we may have banks of frozen cells, tissues, and even organs for emergency use. It will not be necessary to depend upon willing living donors for these transplants. Tissues and organs from persons recently dead, especially after death by accident of persons in their younger years, will provide an ample supply for surgical use.

Biomedical engineering is now developing many kinds of artificial organs, such as hearts, kidneys, and blood vessels, which can be safely used to replace the worn-out or diseased originals. The old man or woman of tomorrow may live on, in better health, for several decades with one or more artificial parts. Much public interest attaches to preliminary, imperfect efforts of this kind, such as the implantation of an artificial heart in Barney Clark in 1982.

Surgery extending to the prenatal life is already a rapidly developing art. The embryo has a remarkable capacity to regenerate lost parts, a capacity which is lost as the fetus grows older. Also, the immunogenetic reactions are weak or absent at first and usually develop late in fetal life or even after birth.

Consequently, the replacement of defective organs might be made much more easily, and with better results in the sequel, were surgery to be undertaken during prenatal life instead of after birth.

BRAVE NEW WORLD

To a geneticist concerned with human heredity, one of the most remarkable achievements of scientific man in the twentieth century has been the development of our understanding of how the invisible genes contained in the tiny human sperms and eggs, that form the only direct biological connection between one generation and the next, establish the nature and the potentialities of each individual. In our forty-six chromosomes we each carry two sets of genes composed of deoxyribose nucleic acid (DNA) molecules. One set is inherited from the mother, the other from the father. In all, there may be about 20,000 to 40,000 of these genes (10,000 to 20,000 per set), which endow us with the ability to synthesize that same number of protein molecules of different sorts and to control their activity by turning them on or off in particular cells and tissues as development proceeds. Some of these proteins are enzymes that control the various chemical processes of growth, development, and maintenance; others are the structural proteins (such as hemoglobin in the blood cells, myosin in the muscle cells, and collagen in the connective tissue) that enable us to function.

All our differences from other animals, indeed all our inherited differences from other human beings, reside in differences in the nature of our proteins and, therefore, of our genes. Only the genes can reproduce themselves, and they do so every time each cell divides. The science of genetics, child of the twentieth century, is the science that relates to the study of these genes and what they do, as well as how they are transmitted in reproduction. All of the modern science of evolution revolves around a knowledge of how these genes mutate, what forces lead to the selection and perpetuation, over the generations, of those forms of genes that endow their possessors with superior capacity to live to reproductive age and to have offspring, as well as how the frequencies of alternative kinds of genes are influenced by chance and by the migrations and interbreeding of more or less isolated populations. Knowledge has progressed from a few basic laws of the inheritance of genes, laws known in 1900, to a complete picture today of the genetic code that relates the chemical sequences of four kinds of nucleotides in the gene to the sequences of amino acids in each protein that is being synthesized.

Mankind is now in almost full possession of the means to modify the hereditary nature of every plant and animal and microbe it wishes. Rapid strides have been made in the isolation and chemical synthesis of genes. Genetic engineering, using enzymes that cut genes out of their residence in the chromosomes and others that splice them into new locations, together with bacterial plasmids or viruses that can be used to introduce the isolated or freshly synthesized genes into bacterial hosts or into human or animal cells grown in culture, offer exciting new possibilities. One of these is to use bacteria as synthesizers of scarce gene products, such as insulin or interferon or many others. Another, still unrealized, is to inoculate the tissues of a person who has a particular genetic defect, with human cells compatible with growth and capable of making the product lacking in the genetically defective person.

If the measure corrects the defect only in the somatic organ most involved, the normal biochemistry may be restored. If, however, the representatives of the defective gene residing in the reproductive cells are not also replaced, reproduction by the person at a later date may transmit the defective gene. Then, depending upon its dominance or recessivity, there will be required, either at once or at some remote later time in a descendant, a fresh artificial gene transfer. In this way the human geneticist would multiply in the population the effects already causing increasing concern because of the reproduction of genetically defective persons whose symptoms have been alleviated by medical treatment. The list of such conditions is impressive in length. It includes such conditions as myopia, sickle cell anemia, schizophrenia, and many lesser known conditions. Rather than so greatly to increase the frequency of defective genes in the population, it would seem better to restrict the reproduction of such affected persons whenever the defect is judged sufficiently severe. But who is to make that decision? What panels of physicians and geneticists, psychologists, and sociologists would possess sufficient wisdom or dare assume such authority?

Indeed, there is nothing to prevent us from meddling with our own hereditary makeup in less desirable ways. This staggering power over the nature of human evolution provokes another great crisis in human affairs—the crisis of values and goals, as I prefer to call it. For the problem that faces us is not so much the problem of *how* to change man's nature, as the problem of *what kind* of human being we wish to populate the earth. Should we breed for uniform superiority of physique, or health, or intellect? Or should we breed for the maintenance of human diversity? Is it possible that superiority in certain respects is incompatible with superiority in certain other ways? Should we produce castes for different callings? Should we regulate and impose social ends on the personal rights of reproduction?

Complete control by individuals and by society over reproduction opens up certain eugenic possibilities like those of Aldous Huxley's *Brave New World*.[5] First, let us recognize that, under our present application of ethics to medical practice, human society is already doing itself a considerable eugenic injury. When an infant or child with some genetic defect—among the 3,000 or more already known—is kept alive by medical means, and its defect is controlled or even eliminated, the child grows to adulthood. The most common sequel is marriage and parenthood. In the past, under a more cruel rule of nature, such persons never lived to reproduce, and their defective genes were thus eliminated from the population. To a certain extent, doctors are only making more work for themselves, for since they have not removed the cause of the defect—the gene—but have only corrected symptoms, the defective gene, when transmitted, does the same harm once again.

To some extent, this dilemma is offset by the fact that mutant genes that in a natural environment would have been defective are rendered innocuous, or nearly so, by the structure of our improved, highly artificial, environment. For example, myopia may once have been a serious handicap, in a state of culture when all depended upon hunting for food. But when

[5]Aldous Huxley, *Brave New World* (New York: Harper & Bros., 1932).

agriculture developed and food was more abundant and to produce it no longer required keen eyesight at a distance, nearsighted persons could find useful employment in various ways. Very much later, when spectacles were invented, myopia almost ceased to be a problem—it shrank to the level of a mere inconvenience. The same trend has happened in respect to many genetic disorders for which there now exist simple remedies: drugs to control diabetes, allergies, gout, and to lessen pain of temporary sorts. Even so, we should keep in mind that it is our environment, technologically maintained, that has lessened the burden of these defective genes, which are now probably far more common than when they caused severe adverse selection against them. Any serious disruption of our technological industrial society would have disastrous effects upon the genetically defective persons in the population—and that is in fact practically everyone, in some respect or other.

Geneticists are looking forward to the day when they can practice genetic surgery, that is, really reach in and transform a defective gene and make it functional again. That will not be easy. If the defect is bodywide, the transformation would have to be done either in the very young embryo, before the cells have begun to multiply, or in the reproductive cells that actually function in making new individuals. First, then, we must perfect the means of culturing the embryos and the reproductive cells in the laboratory. Even so, it will probably always remain easier and simpler to discard defective reproductive cells and to select others from banked or cultured material that is free of known defect. On the other hand, if the defect is limited to the physiological activity of a limited part of the body, such as a single organ, then genetic engineering might be capable of producing a culture of the deficient tissue into which an effective gene might be introduced from a normal donor. The transformed cells could then be injected into the defective organ, or be surgically implanted in it, and successful restoration of the impaired function might take place. This achievement does not look beyond the realm of possibility in the next twenty years.

We now come to the most difficult aspect of the crisis of values and goals. How can one select good strains of reproductive cells? If the same material is used to produce a great many embryos, which are reared into babies, they will be too much alike, like members of the same caste in *Brave New World*. This result might be avoided by never using a single line of reproductive cells more than a few times. There is another, even more serious, difficulty. Nearly all of us carry some defective genes. The average is probably around eight. We lack visible signs of defect because most such defective genes are recessive, that is, they must be present in a double dose, being inherited from both father and mother, in order to produce an evident defect. As long as we have one efficient working gene belonging to any particular pair of genes, enough of the protein it controls is usually made to satisfy general needs. Close relatives, however, have a greatly heightened probability of carrying the same defective gene because of possessing a common ancestor. It would, therefore, be necessary to have strict rules to prevent offspring being produced by persons who were derived from the same lines of banked or cultured reproductive cells, and careful records on the lineage of each person would need to be kept.

For any program of positive selection or genetic manipulation we must clearly establish goals and standards. Difficult, yet easiest, would be to determine that a certain allele A′ was inferior in quality to its alternative, allele A. Then we might substitute allele A for allele A′ either euphenically, by introducing the effective gene into the tissues of the recipient whose own allele is A′ or eugenically, by substituting A for A′ in the same individual's reproductive cells. It has not been demonstrated thus far that the latter transformation is feasible, but conceivably a harmless virus carrier might transduce immature sperm cells or ova as readily as cells of liver or kidney. Let us concede that if this could be done in the case of a sufferer from retinoblastoma, the action would be laudable. Retinoblastoma is a single-gene dominant genetic malignancy, incurable unless stopped short of its passage to the brain by surgical removal of the affected eye, and it is expected to be transmitted to half of the survivor's children. But for the overwhelming majority of genetically conditioned traits no such simple solution is possible. Clearly, for the most significant traits, the values and goals are not for science alone to impose—we are concerned with social values, and which of these, we must ask, is preeminent?

I once suggested that we might agree upon such goals as "freedom from gross physical or mental defects, sound health, high intelligence, general adaptability, integrity of character, and nobility of spirit." I did not imply that these characteristics were in any case fully or even partially genetic in nature. H.J. Muller selected a different list: "genuine warmth of fellow feeling and a cooperative disposition, a depth and breadth of intellectual capacity, moral courage and integrity, an appreciation of nature and art, and an aptness of expression and of communication"; on the physical side, "to better the genetic foundations of health, vigor, and longevity; to reduce the need for sleep; to bring the induction of sedation and stimulation under more voluntary control; and to develop increasing physical tolerances and aptitudes in general."[6] I suspect that any other scientist would select a different, though overlapping, list.

Such qualities cannot be selected without our having ways and means of defining them precisely and measuring them at least roughly in a quantitative way. Obviously the psychologist and the sociologist will need to do a great deal of preliminary work before genetic analysis and understanding of these traits become possible. At present, the genetics of human behavior is still a young, rather undeveloped field.

As for the techniques which are developing so rapidly, their progress is restrained chiefly by the rather small number of biological scientists who have so far entered these fields. Only a handful of workers, for example, are engaged in any study of the histocompatibility genes. These are the very important genes that determine the strong reactions of human tissues against the implantation of foreign cells of different genotype. The studies made by these workers, first with mice and now extended to human cells, indicate that a rather small number of genes, probably less than twelve, are involved. The identification of the specific proteins synthesized under the control of these genes may soon lead to their isolation. It is not too far-fetched to expect that soon it will become possible to inject any fetus or newborn child, prior to the development of the immune reactions, with an appropriate selection of those

[6]H.J. Muller, in *Evolution and Man's Progress*, H. Hoagland and R.W. Burhoe, eds. (New York: Columbia University Press, 1962), pp. 35, 37.

histocompatibility proteins which are foreign to its native genotype. The result would be acceptance of these as if they were a part of "self," just as in the case of twin cattle or other twin mammals who have exchanged cells through a fusion of their placental circulation prior to birth. It would then become possible for anyone so treated to receive with impunity a transfusion, implant of bone marrow, or transplant of any organ from any donor with these same proteins. The whole vexing difficulty of the reactions of recipients of transplants would thus be circumvented.

We may, however, have thus only multiplied our ethical difficulties. The supply of organs available for donation would be even less sufficient than now. The difficult question of who is to have a prior right to an available transplant would be more difficult still if every person were an equally suitable recipient and demand far outstripped supply, even though every person might be an acceptable donor.

If defective genes could be altered in the reproductive cells themselves and restored to normality, that would be better. Short of treatment of reproductive cells grown in culture and followed by fertilization *in vitro* and implantation of the embryo in a surrogate mother, the means are not apparent. Great strides are now being made in the development of just these techniques, however.

As long ago as 1970, piglets conceived in Canada were flown across the Atlantic Ocean in a tube and implanted in a sow of different breed. The parents were three sows and a boar of the Canadian Yorkshire breed, and the embryos were conceived on March 16 at a Canadian Department of Agriculture farm at Hull, Quebec. The foster-mother, at Weybridge, Surrey, in due time delivered her litter of "improved" piglets. It seems quite clear that the techniques I am describing can be applied to mankind, after being improved in minor ways.

Human ova obtained after hormonal treatment to induce multiple ovulation have been fertilized in the laboratory, and the embryos have been raised to the blastocyst stage when implantation normally occurs. The births of successfully implanted human embryos as normal babes, from 1978 on, in England, Australia, the United States, and perhaps in India, have made worldwide news. So far, the new reproductive technique has been applied only to enable women whose oviducts are blocked to have their own children, but there is no technical impediment to the implantation of viable embryos in surrogate mothers, except for possible legal and ethical complications. Inasmuch as these differ from society to society, we may reasonably expect that in some countries surrogate motherhood may become a common practice. Mouse embryos are today put in the "deep-freeze" at the temperature of liquid nitrogen and have been thawed and implanted in surrogate mothers many years later, to develop quite normally. There seems to be no reason why they could not be kept in the frozen, inanimate condition for a century or more. The human embryo could be treated similarly, for biologically it responds in the same manner. Thus it would become possible to bank human embryos in deep underground laboratories where no nuclear explosions could disturb them, to be safely implanted for the perpetuation of the population after a holocaust above ground. The banking of human sperms for reproductive purposes will doubtless be supplemented safely in the future by the banking of human ova and of frozen embryos, as we do now in breeding cattle, sheep, and horses.

In the mouse, a technique has now been developed for securing implantation in an artificial membrane composed of precipitated collagen, and development has progressed to the early stages of formation of a placenta. I fully expect that another decade of research, even without any increase in the small number of workers actively engaged in such investigations, will bring us to a stage when human "prenatal adoptions" can be routinely performed, and perhaps an artificial placenta may have succeeded in supplying the needs of a mammalian embryo to full term.

In these ways, the reproductive cells of selected individuals might be utilized, even long after their deaths, to produce in the laboratory embryos that might be implanted in the womb of a surrogate mother, or even grown in artificial cultures. Techniques will very probably be developed for the cultivation in the laboratory of portions of human ovary and testis, a procedure that will permit successful continuous production of mature ova or sperms. A report of success in the production of mature ova from cultured mouse ovaries leads one to expect that sufficient persistence by skilled biologists can attain the successful cultivation of human reproductive organs, continuous production of eggs and sperms, and formation by fertilization in the laboratory of as many human embryos as might be wished.

I am frequently asked why anyone should wish to pursue this goal. "Aren't the age-old ways of making babies good enough?" one questions. Several reasons may be given why such exploration of new possibilities is desirable. Only by studying the development of the human embryo and fetus under continuous observation and under various conditions can medical scientists really learn what factors produce particular kinds of abnormalities and how these may be corrected or avoided. Moreover, the practice of "prenatal adoption," that is, the implantation of a healthy selected embryo in a foster mother's womb, appears to have fewer religious and legal objections than the present practice of artificial insemination of a woman by sperm other than her husband's, whether with or without her husband's consent. The development of the implanted fetus within the mother and its normal delivery at full term may be expected to engender the maternal and paternal feelings of the "parents" far more fully than can the adoption of a child already several years old. Moreover, most couples who are sterile may in this way have all the experiences of parenthood.

There are other reasons why such practices might be adopted, if not in the United States, then possibly in other countries. Banking of reproductive cells taken from individuals around twenty years of age may serve to prevent the accumulation of detrimental mutations that normally occurs with advancing age. From the standpoint of eugenics, parents should have their children while they are young. From the standpoint of economics and population control, marriage is often long postponed, and children are produced by parents in their thirties and forties. Healthy children may of course be produced by parents even at an advanced age, but the odds are not as good. By banking reproductive cells under conditions where mutation is reduced to the lowest level, and then using implantation of an embryo produced by artificial fertilization, older parents may have children as free of defect, on the average, as when they were young. Persons such as astronauts who venture into especially hazardous environments, where they are very likely to be exposed to sizable doses of high-energy radiations, might be similarly well-advised to bank their reproductive cells under safe conditions.

Another reason, to be discussed in the final chapter of this book, is to protect the survival of the population in the event of a nuclear war. The only meaningful civil defense is to conduct such experimentation deep underground, in a city of laboratories hewn from solid rock, where the seeds of all necessary future life—of our crop plants and domestic animals and other species that could constitute a viable biome, an ecological whole—might be saved from destruction until aboveground radiation had returned to tolerable levels.

The artificial regulation of conception, pregnancy, and birth are of course necessary not only for such emergencies, but also to prevent the imminent overpopulation of the world. Modern methods of contraception may come to include the implantation under the skin of a capsule of hormone that will, in the female, prevent ovulation, or in the male prevent the maturation of spermatozoa, as microdoses seep into the circulation over a period of months. Such methods will make our current reliance for contraception on the "pill" or the intrauterine device, with their low but definite dangers, seem truly antiquated. Advance through research in this field will automatically make induced abortion a thing of the past.

We might even predict that, by the year 2000, many countries will have reached such a population density in relation to their food supply that no further increase can be tolerated. A marriage certificate might then bear two coupons entitling the couple to produce two children, no more.[7] Restrictive tax measures, such as an income tax graduated more heavily as the number of children increases, or even temporary sterilization by court order might be utilized by some countries in desperation. Temporary sterilization of the male or female by implantation of a hormone capsule might be sufficient.

In the last two decades we have had a considerable revolution in America in sexual behavior. The widening use of contraceptives and the availability of legal abortion upon demand

[7]This facetious proposal has been made independently more than once, in particular by Kenneth Boulding, in *The Meaning of the Twentieth Century* (New York: Harper & Row, 1964).

have converged to isolate sexual behavior from its fundamental biological function leading to reproduction. The further severance of reproduction under artificial control from any relation to the sex life may have grave but unpredictable effects upon the stability of the family. Robert Morison has pointed out the existence of this grave threat to the continuance of the family as the basic social unit.[8] Can we safely, after a million years of human and prehuman evolution during which the family has been the basis of all protection of the young, their education, and their nurture, give it up? What might be the psychological consequences of a population with no personal ties either to the younger or the older generation. Can we look forward to a brotherhood of man when there are no more parents, brothers, or children, only unrelated persons?

Prometheus and Pandora are truly inseparable. We must begin to apply more consistently and conscientiously to all of our professional efforts the lesson of our times, that technological developments, which at the time seem highly beneficial and moral, may be seen, in the light of their long-range effects and side effects, to be detrimental and immoral. No profession can henceforth escape this responsibility. As Emilio Daddario has written, in a definition of "technology assessment," we must "identify irreversible trends toward catastrophic incompatibility."[9] Now that the work of the physician is directed more fully toward the constitutional disorders, the congenital defects, and the errors of metabolism, all of which are so largely of genetic etiology, the partnership of physician and geneticist must, as it should, become far closer. I suggest the wisdom of a much greater training and involvement of the geneticist in medicine, and likewise a much greater training and involvement of the physician in genetics. Perhaps each of us can counteract the astigmatism of the other.

At the bottom of Pandora's box was hope, which Prometheus claimed as one of his gifts to man. Yet it is not "*blind* hope" that we want or need, but a clear-sighted hope that analyzes each system fully: all its input and consequences, the offsetting benefits and detriments. The ethical and moral, as

[8]Robert S. Morison, "Where is biology taking us?" *Science* 155(1967):429–33.

[9]Emilio Q. Daddario, "Priorities for the future: Guidelines for technology assessment," in *1971 Britannica Yearbook of Science and the Future*, p. 310.

well as the physical and the social. The collective, as well as the individual. We live in a world of conflicting human values. The preservation and multiplication of the molecule may be at the expense of the life of the cell. The multiplication of the cell can clearly become uncontrolled and malignant at the expense of the life of the person. The rights of the individual may conflict with the good of the state or of society. As we are one body, so are we " one world." When I read in the Bill of Human Rights of the United Nations that one incorrigible right of the individual is to reproduce, and that the right of every person to have a family is a basic human right that must not be infringed, I wonder whether this "right" is indeed to remain unrestricted. Is it not equally a right of every person to be born physically and mentally sound, capable of developing fully into a mature individual? This right, it seems to me, is paramount. It is one that both physician and geneticist can work to guarantee.

Our joint efforts must not stop at that point. The unbridled growth of the world's populations is our problem. The increasing concentration of chemical mutagens and toxic wastes in man's environment is our problem. The growing exposures to high-energy radiation are our problem. Social security, to say nothing of the delivery to all of adequate health care—these too are our problems. If indeed we are to retain *clear-sighted* hope for the future of mankind, the geneticist must work toward the optimal environment for each and all. The physician must work toward the optimal health of every person and collectively of every group, a goal that again implies the provision of an optimal environment for each and all. The ecologist must work for the preservation of a sound environment and the protection of all our natural resources. Every citizen must become aware of the peril, as well as the promise, of our future world.

Part II
The Nature of Biological Science

2 Explanation in Biology

It does not seem profitable to regard scientific *description* as constituting a form of explanation. By *explanation* we are to understand an interpretation of phenomena involving "cause" in a scientific sense. Yet cause is not to be conceived in a simplistic, unitary sense. As Bertrand Russell has pointed out, nothing less than the total prior state of the universe is causal for any particular event.[1] Nevertheless, in making scientific hypotheses upon which predictions may be based and subjected to test, certain recurrent, regularly associated relationships are usually sought out, on the basis of which one may predict from the occurrence of the earlier one that the other will follow with a certain probability. This relationship, to be sure, does not distinguish between "cause" and "correlated event"—between the phenomena in direct sequential relationship and those on different paths branching from the same initial event. I have elsewhere discussed this ambiguity and criticized the identification of "necessary connexion" (that is, cause—according to Hume, Mach, Pearson, Braithwaite, and others) with "constant conjunction."[2] Unless no interval of time at all intervenes

This essay originated as a contribution to a symposium on the subject of "Explanation in Biology," held at Asilomar, California in June 1968. The entire symposium was printed in the *Journal of the History of Biology*, vol. 2, no. 1 (1969). The present essay is reprinted by permission.

[1]Bertrand Russell, "On the Notion of Cause, with Applications to the Free-Will Problem," in *Readings in the Philosophy of Science*, H. Feigl and May Brodbeck, eds. (New York: Appleton-Century-Crofts, 1953), p. 392.

[2]Bentley Glass, "The Relation of the Physical Sciences to Biology—Indeterminacy and Causality," in *Philosophy of Science. The Delaware Seminar*, vol. I, 1961-62, B. Baumrin, ed. (New York and London: John Wiley

between A (cause) and B (effect), an assumption which Russell points out is untenable, and unless therefore no intermediate state could possibly exist, correlated events may exhibit perfect conjunction and temporal sequence, yet belong to separate branches of the nexus of events. In the simplistic view assumed by those who have followed Hume in adopting constant connection as the criterion of cause, only the simple chain reaction, with one event at each interval of time, is allowed. This view may appeal to the physical scientist, but the biologist and biochemist are too familiar with branching step reactions to accept such a view.

Verbal explanations have pervaded biology in the past, to its real detriment. Thus, the vague and untestable concept of protoplasm as the material substance of life long prevented a critical analysis of living systems. The theory of the gene, so fruitful in its early history, became a deadening influence because of general lack of sufficient appreciation of the distinction between the operational modes (mutation, recombination, function) whereby a given gene becomes known. The conclusion by Louis J. Stadler in 1954 that geneticists must clarify their concepts through sharper use of operationally defined concepts,[3] and the somewhat later brilliant work of Seymour Benzer[4] carrying out that injunction, did much to pave the way for the great revolution in genetic thinking that took place in the 1950s and marked the advent of the era of the genetic code.

Biology is the science of life, but what "life" itself is we cannot define in precise operational terms. As William S. Beck has stated, "Life and nonlife... are ends of a spectrum whose graded quantity is complexity: life is on the complex end, nonlife on the simple. Between the two is a middle ground which is neither one nor the other. It is, one might say, what it is."[5] That view is further emphasized in the title of an essay written by N.W. Pirie, "The Meaninglessness of the Terms of

and Sons, 1963), pp. 223–57; see especially pp. 233–39. (See Chapter 3 in the present volume.)

[3]L.J. Stadler, "The Gene," *Science* 120 (1954): 811–19.

[4]Seymour Benzer, "The Elementary Units of Heredity," in *A Symposium on the Chemical Basis of Heredity*, W.D. McElroy and B. Glass, eds. (Baltimore: The Johns Hopkins Press, 1957), pp. 70–93.

[5]W.S. Beck, *Modern Science and the Nature of Life* (New York: Harcourt, Brace, 1957), p. 186.

'Life' and 'Living.'"[6] Granting the foregoing, explanation in biology must clearly grade into the modes of scientific explanation used in the physical sciences. The question remains: are there other modes of explanation, either current or outmoded, which are peculiar to biology, either because of its history as a science or because of the nature of the phenomena and interrelationships at the complex end of the spectrum of nature?

I believe that there are. We may well start with Aristotle's doctrine of the Four Causes, since Aristotle was the greatest biologist of the ancient world and rather surely formed his epistemology on the basis of his studies of living creatures and their development. Aristotle distinguished four categories of cause: the *Material Cause,* or matter, out of which the thing is made; the *Motive,* or *Efficient, Cause,* which sets the process going and is that *by which* the thing is made; the *Formal Cause,* or form, which is responsible for the character of the course which the process follows, and is the special character of the development that the process pursues; and the *Final Cause,* the end or object toward which a formative process advances, its *logos,* its rational purpose, the complete and mature final stage. These definitions are taken from the excellent discussion by A.L. Peck in the Introduction to his translation of Aristotle's *Generation of Animals.*[7] Note further that, in terms of the development of a chicken or a dog, the Material Cause was conceived by Aristotle to be supplied by the female parent, the Motive or Efficient Cause by the male parent. The Formal Cause consisted of the sequence of stages through which the embryo and young chick or puppy pass during the course of its development, and the Final Cause is represented by the mature bird or dog, the goal of the developmental process.

The modern biologist has translated these varieties of cause into his explanations of biological phenomena. The Material Cause has become the molecular explanation of the biochemist

[6]N.W. Pirie, in *Perspectives in Biochemistry,* J. Needham and D.E. Green, eds. (Cambridge, England: University Press, 1939), pp. 11–22.

[7]Aristotle, *Generation of Animals,* A.L. Peck, trans. (Cambridge, Mass.: Loeb Classical Library, Harvard University Press, 1943), pp. xxxvii–lxiv. At the Asilomar Conference my attention was directed to a similar identification of Aristotle's Four Causes with modern biological conceptions of cause. Unknown to me, the same idea had previously been set forth by John Herman Randall, Jr., in *Aristotle* (New York: Columbia University Press, 1960).

and geneticist. In terms of metabolic maps, each substrate of a step in a chain of reactions is the "material cause" of the product or products. Thus, in the step reaction A → B → C, A is the material cause of B; B the material cause of C. When, more often, two or more substrates react and enter into the formation of one or more products, each one must be regarded as a "material cause." Not only is the immediate substrate in such relation to the product, but any material substance at an earlier step which becomes transformed into the product is so to be regarded. Thus, in the well-known steps of glycolysis (the digestive breakdown of glucose, ordinary corn sugar), glucose, glucose-6-phosphate, fructose-6-phosphate, fructose-1,6-diphosphate, glyceraldehyde-3-phosphate, 1,3-diphosphoglyceric acid, 3-phosphoglyceric acid, 2-phosphoglyceric acid, and phosphoenolpyruvic acid are all precursors and principal substrates for the formation of pyruvic acid, in the order named. Some of the steps simply involve rearrangements of the substrate molecule; others require the addition or removal of phosphate groups, water or hydrogen atoms. The small chemical groups (such as the phosphate group or water) may be regarded as material causes no less than the great ones. Many biological explanations stop at this molecular level. It is purely descriptive except for the introduction of the dimension of time.

The Efficient Cause, in modern terms, relates to the energetics of the biological process: thermodynamics, the kinetics of reactions, and the effects of catalytic agents, especially enzymes, upon the required activation energies and upon the steric placement of substrate molecules that so greatly enhances reactivity. Since most energy is supplied in biological systems in the form of chemical energy, the boundary between this category of causes and the first is none too sharp. Nevertheless, I would place here those biological explanations that relate to the genetic control of protein synthesis, the enzymatic control of metabolic reactions, the capture of solar energy through photosynthesis, chemosynthetic mechanisms, and all positive and negative feedback regulatory mechanisms that promote homeostasis. Life has been characterized as an eddy in the downward flow of energy to the sea of entropy, a good analogy calling attention to the evanescent storage of energy and the reversal of the Second Law of Thermodynamics that takes place in living organisms.[8] An explanation of a sort, perhaps! Certainly many biologists are satisfied to leave their explanations of the phenomena they study at this level. Trace the energy to adenosine triphosphate (ATP), the reaction to the presence or absence of an enzyme, the developing characteristic to a gene—and they feel the job is sufficiently done.

The Formal Cause in modern biology takes the shape of explanations that relate form to function. The complementarity of form and function is a basic biological conception with little counterpart in the physical sciences, where the idea of function—the performance of action—directed toward an appropriate adjustment to the environment—is essentially lacking. The idea of function is, in fact, a teleological notion unless construed in the light of evolutionary adaptation as a product of natural selection. The "Wisdom of Divine Providence," seen in the eighteenth and nineteenth centuries to be manifestly revealed in the relation of structure to function in the human hand or in the mutual suitability of bees and flowers for pollination, is witness to the essentially teleological bias relating the Formal Cause to the Final Cause. Darwin freed us from this constraint by showing clearly just how natural processes can produce "appropriate" structure and function without divine intervention. The same arguments apply, furthermore, to the appropriateness of successive stages in the development of an organism. Natural selection has worked on all stages of the life cycle, not on the adult stages alone. The apparent foresight with which each stage of development prepares for the advent of the next one and looks forward to the goal of a mature, reproducing adult is actually a relation of form and function that looks back to the programmed instructions for the developmental process, instructions located in the genotype of the individual. The genotype is well-adapted to the usual environment in which development takes place because of the natural selection of genotypes that occurs in every generation—

[8]The idea of life as a whirlpool ("tourbillon continuel") of matter entering and leaving the organism was a favorite of Cuvier, as cited and quoted by Lawrence K. Henderson in *The Fitness of the Environment* (New York: Macmillan, 1927), pp. 23–34. The subsequent relation to the Second Law of Thermodynamics was pointed out by Henderson himself, as well as by many others; see in particular Homer W. Smith, *Kamongo* (New York: Viking Press, 1932), pp. 147–54.

that perpetuation of the better adapted and elimination of the less fit in a world in which there is differential mortality and fertility of genotypes.

The Final Cause, the teleological aspect of Aristotle's view of nature, is replaced in modern biology by explanations of a teleonomic sort, to use the term usefully introduced by C.S. Pittendrigh.[9] The teleonomic explanation explains adaptations of form and function, as well as of genotype and final adult characteristics, on the basis of natural selection. On this view it is the environment that serves as the screening element in the perpetuation of whatever mutant forms of genes compete in any given population. The environment in this context is to be understood in the widest possible sense. It operates at different levels of the living world. At the molecular level, the enzyme or the messenger RNA (ribose nucleic acid) may be regarded as the environmental pattern that affects the reactivity of substrates or the synthesis of proteins. The special structures, such as mitochondria, in which enzymes of the respiratory chain are arrayed in assembly lines, provide a further example of the ordered molecular environment whereby reactions can be carried out locally that would never be perceptible in a randomized system composed of the same constituents. The environment of the individual living cell in a multicellular organism is comprised of the other cells of the tissue and of the intercellular fluids. Natural selection must have served also to evolve the mutual fitness of cells, both those of the same sort in a tissue and those of different tissues making up a single organ and performing a joint function. To natural selection is also due the remarkable fitness and constancy of the internal milieu at which physiologists since Claude Bernard have consistently marvelled, and upon which the homeostasis of the organism and its regulation by neural and hormonal messages largely depend. The individual is of course molded both temporarily and genetically by the pressures of the external environment, which includes especially the influences of other beings of the same population and species, as well as members of other living species in the community of which the individual is part.

[9]C.S. Pittendrigh, "Adaptation, Natural Selection, and Behavior," in *Behavior and Evolution*, A. Roe and G.G. Simpson, eds. (New Haven: Yale University Press, 1958), p. 394.

At all these levels of interplay between living system and environment, the teleonomic, evolutionary explanation is central. Life on this earth has had a history, to be understood only in terms of the evolutionary process. The fitness of living forms—all their remarkable adaptations—we attribute to the repeated action of natural selection through times past on individuals, their characteristics, their genotypes, and ultimately, their individual genes. Many times in this history the thing that actually did happen was but one of many possibilities, The entire history thus adds up to a generation of highly improbable creatures in the multitude of possible types never realized, at least on this planet. There is thus a uniqueness about the evolutionary history of life that escapes prediction although it is fully explicable on the basis of the current theory of mutation, natural selection, and isolating barriers as the essential factors in the origin of species.

The types of biological explanations chiefly employed at the present stage of the science may be summarized in the form of a series of questions that relate one aspect of life to another. Another biologist might suggest others. The following list is my own.

1. How is a specific character related to the genotype?
   a. How is a specific protein determined by the genetic code?
   b. How is a specific character related to the function of a particular protein?
2. How is one stage of development and life cycle related to those that precede and those that succeed it?
   a. How is form generated?
   b. How are the definitive features of a species replicated in successive generations?
3. How is one level of organization related to those below it and above it?
   a. What special laws obtain at particular levels? Are they reducible to the laws operating at lower levels?
4. How is the living system related to its environment at each level of organization?
5. How is form related to function in each specific case?

6. How are regulation and homeostasis achieved at each level from the molecular to the biome?
7. How is specific response related to specific stimulus over short and long intervals of time?
8. How do living organisms acquire adaptations through the evolutionary process?

Each of these questions might be explicated on a simple descriptive basis; each one can also be pursued through all the levels of biological organization. The molecular biologist, the ecologist, the developmental biologist, and the evolutionist will give different, but complementary, answers to the same question. Only the total matrix of biological explanation can fully suffice.

3 The Relation of the Physical Sciences to Biology

One might explore the relations of the physical sciences to the life sciences by surveying the deepening extent of the dependence of biologists upon physical and chemical techniques and the remarkable growth in our time of molecular biology. To do that would, I believe, have less interest (for philosophers of science) than to consider the current status of a much-debated question that has long troubled, and even now intensely concerns, biologists—the question of causality, the conflict of determinism and purposiveness, of the concept of the organism as machine with the concept of the organism as a system in which the whole is more than the sum of the properties and actions of independent components. The question may be posed succinctly in the following form: *Do the laws of physical science comprehend all the laws of biology?*

There is frequently semantic confusion in the discussion of this question. If physics and chemistry be defined as the realm of matter and energy, then by definition all biological phenomena are embraced within physics and chemistry, since no scientific observations can be made except of matter and the changes in matter attributed to physical forces. It may be helpful to rephrase the question in several alternative forms, as follows:

This essay originated as a contribution to the Delaware Seminar in the Philosophy of Science and was published in volume I of that series (New York: Interscience Publishers, John Wiley & Sons, 1963), pp. 223–57, edited by Bernard Baumrin. It is reprinted here in revised form with the permission of the publishers and the University of Delaware Press.

1. Do the laws of physics and chemistry which are demonstrable in nonliving systems also hold true in living systems? To this question the modern biologist gives an unqualified yes.

2. Do the laws of physics and chemistry which are demonstrable in nonliving systems constitute all the presently demonstrable laws of living systems? To this question the answer is an equally unqualified no.

3. Can the laws of physical science which are demonstrable in nonliving systems be expected ultimately to explain all the laws of living systems? This is the aspect of the question where differences of opinion arise. But the reasons given are often quite different and require scrutiny. Those who would give an affirmative answer to question 3 point to the steady extension of physical law in the realm of biology as research proceeds, and they maintain that since problems once held to be inexplicable in physicochemical terms can in many instances now be explained quite satisfactorily in such terms, it is only a matter of time until all biological phenomena will be embraced by and be explicable in terms of physical and chemical laws. This prediction is, of course, based simply on the extrapolation of a persistent trend. On the other hand, those who give a negative reply to question 3 do so for a variety of reasons, some of which are clearly based more on faith than on reason.

FOUR BASES OF DISTINCTION BETWEEN BIOLOGY AND THE PHYSICAL SCIENCES

In *The Structure of Science,* chapter 12, Ernest Nagel has discussed four arguments advanced by some biologists to support the belief that biology inherently cannot be reduced to physicochemical terms.[1] The first of these is simply the claim that biology deals with characteristics and laws that do not occur among nonliving systems and that the techniques of observation and analysis are often very different. At first sight this argument seems entirely circular: the claim assumes the

[1]Ernest Nagel, *The Structure of Science: Problems in the Logic of Scientific Explanation* (New York: Harcourt, Brace and World, 1961), chap. 12, pp. 398–446.

very point at issue. Nagel rightly disposes of this argument very briefly; and yet we will find, I think, that in one particular form it will come back to trouble us again. It may, in fact, be the most significant of all the arguments presented.

The second argument is not, I believe, sufficiently distinct from the third in Nagel's treatment of the issues. It is, in brief, that biologists must deal not only with *structures,* that is, with organized forms of matter as seen in living systems, but also with *functions.* Functions may be considered in two aspects. On the one hand, a biological function is an organized form of action as seen in living systems. In this sense, it is strictly physicochemical in nature. Thus, in the contraction of muscle there is a chain reaction in which each step is the cause of the next one. The delivery of chemical energy from the high-energy phosphate bonds of adenosine triphosphate molecules to the actin and myosin molecules of a muscle fibril results in the formation of a shorter actomyosin complex and consequently in the shortening of the muscle fibril. This effect is multiplied by repeated occurrence in the numerous fibrils making up a muscle fiber and the numerous fibers making up a muscle, and finally culminates in the appropriate contraction of the muscle that moves certain parts of the skeleton in relation to others. This series of steps seems in no way different from many chain reactions in nonliving systems, such as (1) the transmission of momentum from one billiard ball through a series of others to a final ball which moves away with the speed and trajectory of the first—now motionless—ball; or (2) the transformation of mechanical vibrations into electrical variations, the amplification of electrical effects in a triode tube or transistor, and the subsequent conversion of electrical variations into sound patterns that reproduce the original sounds which have been frozen into patterns in the grooves of a phonograph disk. Function in this sense is no different in biological and physical systems. It describes the relation of the final term in the causal chain to its antecedent steps.

But there is another aspect of function, its purposive, or teleological, aspect. Nagel discusses quite fully this aspect of function and its dominating position in biological thinking, but, I fear, unsatisfactorily. His argument and explication are limited to the causal, chain-reaction relationship just discussed, and he readily succeeds in showing the similarity of the biological chain of steps to those of the physical sciences. He concludes that the distinction is largely semantic, a matter of emphasis upon the "*consequences* for a given biological system of a constituent part or process" in the case of the teleologic explanation, and of the "*conditions* under which the system persists in its characteristic organization and activities" in the nonteleological formulation of the same system. I believe that a more important distinction relates to the evolutionary origin of the adaptations of living beings, which have always so profoundly impressed the biologist and the philosopher. Is there a valid distinction here, in that, for the most part, nonliving systems do not evolve, are not subject to the increasingly greater adaptiveness that results from the action of natural selection? This problem will be considered more fully later.

The fourth argument considered by Nagel is that of the proponents of some form of holistic view of organisms. This is the view that there are emergent qualities in living systems which disappear upon analysis and dissection of the system into its component parts and which are not simply the products of the interactions of the parts in physical ways. For some biologists this organic whole is the entire organism; for others it is the individual cell.[2] As E.S. Russell[3] has stated, the cardinal principles held by the organismic biologists are as follows: "The activity of the whole cannot be fully explained in terms of the activities of the parts isolated by analysis"; and "No part of any living entity and no single process of any complex organic unity can be fully understood in isolation from the structure and activities of the organism as a whole." Here I believe most biologists of the present time will agree fully with Nagel's conclusion:

> [O]rganismic biologists have not established the absolute autonomy of biology or the inherent impossibility of physicochemical explanations of vital phenomena. Nevertheless, the stress they place on the hierarchical organization of

[2]Barry Commoner, "In Defence of Biology," *Science* 133 (1961): 1745–48.

[3]E.S. Russell, *The Interpretation of Development and Heredity* (Oxford: Oxford University Press, 1930).

> living things and on the mutual dependence of organic parts is not a misplaced one.[4]

The value of the latter, however, relates to a "wise strategy of research" which may be necessary in biology for a long time to come, to the extent to which there are biological sciences that remain inexplicable in terms of *current* physical and chemical theory. But this is not to admit what some of the organismic biologists would claim, that the method of analysis of an entity into its parts is inapplicable or is less valuable in biology than in the physical sciences. Analysis remains the necessary method of science. Treatment of entities as wholes remains entirely descriptive and leaves the organism a "black box." One may carry the argument further. The modern biologist who deals with ecological relationships treats not the individual organism, but the population, the community of different organisms, and the entire ecosystem as parts of the increasing hierarchy of life. Carried to its logical end, the insistence of some organismic biologists on study only of the "whole" would lead only to contemplation of the universe in its full cosmic extent. Yet not even these organismic biologists will deny that consideration of the whole organism, abstracted from its environment, leads to a degree of understanding impossible if one must deal with an entire population, or community, or ecosystem as an indivisible unity.

THE BORDERLINE BETWEEN LIFE AND THE NONLIVING

In 1935 Wendell Stanley succeeded in crystallizing the tobacco mosaic virus. In crystalline form the virus could be bottled and kept inert and "lifeless" indefinitely. Redissolved in water and placed on the surface of a tobacco leaf, the virus particles infected the leaf, used the leaf's stores of organic compounds, and rapidly multiplied. This discovery quickly brought about a great debate among biologists regarding the criteria of life. Was the virus alive or not? Can an organized

[4]Nagel, op. cit., p. 444.

chemical system at one time be lifeless and at another be alive? Or is multiplication of an organized system not only a necessary, but also a sufficient, criterion of life? The discussion was well pointed up in an essay written by N.W. Pirie in 1937 and entitled "The Meaninglessness of the Terms 'Life' and 'Living,'"[5] although by no means everyone would agree with his conclusion that they are indeed meaningless.

During the ensuing nearly half century a great deal of scientific study of viruses has been carried out. The result has been rather surprising. Virus particles are far more complicated in organization than was first thought when it was discovered that they could be crystallized, supposedly a criterion of chemical purity. They consist of protein and nucleic acid, and the electron microscope reveals an elaborate arrangement of parts, including an injection apparatus, at least in some viruses. The conclusion that they represent the borderline between the living and the nonliving holds true; but viruses seem far too complex to have come into existence without an evolutionary process of many steps and a long history. Hence the argument has shifted. On the one hand we find the biochemist Isaac Asimov now saying, in answer to the question, "At what moment in evolution did life appear?" the following: "Then, eventually, must have come the key step—the formation, through chance combination of a nucleic acid molecule, capable of inducing replication. That moment marked the beginning of life."[6] On the opposing side, Barry Commoner replies that it is nonsense to say that "modern science has all but wiped out the borderline between life and non-life," as Asimov claims. On the contrary, says Commoner:

> Since biology is the science of life, any successful obliteration of the distinction between living things and other forms of matter ends forever the usefulness of biology as a separate science. If the foregoing sentence is even remotely correct, biology is not only under attack; it has been annihilated.[7]

[5]N.W. Pirie, "The Meaninglessness of the Terms 'Life' and 'Living,'" in *Perspectives in Biochemistry*, J. Needham and D.E. Green, eds. (Cambridge: Cambridge University Press, 1939), pp. 11–22.

[6]Isaac Asimov, *The Intelligent Man's Guide to Modern Science*, vol. 2: The Biological Sciences (New York: Basic Books, 1960).

[7]Commoner, op. cit.

That remark leaves untouched the question of the line of demarcation. Commoner places this at the level of organization of the cell: "Analysis of living systems, based on modern physical and chemical theory, leads to the conclusion that life is unique and that it cannot be reduced to the property of a single substance or of a system less complex than a living cell."

Many biologists hold that life cannot be defined in terms of a single characteristic property; but all of them, I think, are agreed that reproduction is a necessary characteristic of any form of life that survives more than a single generation! If we look at the reproductive process, it is very clear that it is indeed a cellular phenomenon, that genetic continuity has its basis in the ability of a single cell to grow, divide, and make two cells. Cells lacking the deoxyribose nucleic acid (DNA), which is the hereditary material, are unable to divide. This DNA may be organized into chromosomes, as is usually the case, or dispersed in the cell, as in blue-green algae. Studies of cell division indicate unmistakably that duplication of the DNA precedes, and perhaps triggers, the other complicated events of cell division. Consequently, many biologists have come to think that the capacity of DNA to replicate itself is the ultimate basis of reproduction. (Some viruses contain only RNA, ribose nucleic acid, by means of which they are able to utilize the DNA and protein-synthesizing systems of their hosts. This status only confirms the primary role of DNA in replication and heredity.)

There is, consequently, ground for believing that DNA is the substance possessing the ultimate reproductive power, just as it is demonstrably the hereditary material. Yet one might still say—and this is perhaps the essence of Commoner's argument—that DNA cannot replicate itself except within the very special environment of a living cell. Even the virus DNA must be injected into a living cell before it can find the supply of very special compounds required for its replication. It turns out replicas of itself and sheathes them in proper coats of protein by deforming and misdirecting the normal metabolism of the DNA in the host cell, like a robot tank that might get into a factory designed for making trucks and that somehow succeeded in shifting the controls so that henceforth tanks instead of trucks came off the assembly line.

Yet the argument that the entire living cell is therefore the smallest living unit that can reproduce, the least common denominator of life, has a hollow ring. It strangely reminds us of the century-old argument that organic compounds could never be produced outside of living organisms—an argument quickly shown to be false by the synthesis of urea, and subsequently of many other organic compounds, in the laboratory. Clearly the cell provides the necessary environment for the replication of DNA, but is the entire cell necessary for that? May the biologist not discover just what substrates and enzymes are required and subsequently duplicate in vitro this essential process of life? Frankly, I think we are on the verge of this discovery. Arthur Kornberg's discovery of a system that will synthesize DNA in the test tube is virtually complete.[8] One needs, in this system, the four kinds of nucleotide triphosphates of which the DNA is composed. One also needs an enzyme—and this has been extracted from the common colon bacillus. One needs a chemical source of energy—adenosine triphosphate will serve beautifully. And finally, one must have a model, a small amount of native DNA that will serve as a primer, or template, for the new molecules to copy. Given all these, DNA molecules are synthesized in the test tube abundantly, and are indistinguishable from the kind of DNA molecules that are used as a model. Take the primer from the colon bacillus, and one obtains colon bacillus DNA; take it from the thymus gland of a calf, and one gets bovine DNA. This process is surely what Asimov had in mind when he made his seemingly rash statement. If even the replication of the hereditary, all-controlling DNA can be carried out in a simple nonliving system, is there reason to suppose that eventually one cannot isolate and carry on all other significant living processes *in vitro*? The photosynthesis of green plants, for example, can now be carried on in suspensions of isolated chloroplasts. Respiration and the generation of an abundant supply of ATP can be performed in isolated mitochondria. What is more fundamental to life than these three processes?

[8]Arthur Kornberg, "Pathways of Enzymatic Synthesis of Nucleotides and Polynucleotides," in *The Chemical Basis of Heredity*, W.D. McElroy and Bentley Glass, eds. (Baltimore: Johns Hopkins Press, 1957), pp. 579–608.

If, on the other hand, one says that life is no single state or process, however complex and indispensable to life it may be, but is precisely that complex interrelationship of processes that occurs only in a living cell, one ends not with an explanation but with a tautology. Nor can one be sure that in the end human ingenuity may not produce a system with even that level of complexity of organization. The creation of life from the nonliving might well be the ultimate demonstration of the meaninglessness of the distinction, of the fallacy of the borderline.

There is another aspect of this question which has received much attention from biologists during the past half century. First introduced by Haldane and Oparin,[9] this is the theory of the evolutionary origin of life from the nonliving. Oparin espouses the idea of coacervates, globular droplets of complex protein nature which form when oppositely charged colloidal particles in suspension are mixed. The coacervates grow in size by adsorbing and absorbing materials from the solution about them; they divide into two when they reach a maximum size; they have a boundary, like the membrane of a living cell. Other theorists have championed the view that the nucleic acid molecules came first and then organized about themselves, from the primeval soup, a more concentrated protein environment in which they could replicate themselves. To decide between these theories is very like trying to answer the riddle of the chicken and the egg—which came first? As the biologist believes that egg and chicken must have evolved together, so it seems to me most reasonable to believe that life commenced with the incorporation of DNA into coacervates. For this to have happened, obviously both must have had a prior existence! The important matter is the presence in the world of natural selection from the very beginning of these steps. Oparin's chief originality consists in his suggestion that natural selection antedated "life." Whatever coacervates were most stable, could grow most rapidly, and when they divided most successfully passed on to their progeny their own superior characteristics—these types would soon

[9]J.B.S. Haldane, "The Origin of Life," *Rationalist Annual*, 1928, pp. 148–53; A.I. Oparin, *The Origin of Life on the Earth*, 3d ed., trans. by Ann Synge (New York: Academic Press, 1957).

outnumber the less stable, less rapidly growing and dividing, less hereditary forms. Moreover, in a situation where the supply of organic compounds formed by long ages of random molecular changes under the influence of heat and radiation must have been very scanty and very dilute, there would be a strong premium upon these qualities. Soon the supply of organic compounds would be cornered by the better endowed forms, for as the others dissolved, their materials would rapidly be taken over by the former. Predatory attacks on other "living" forms would introduce a more active struggle for existence. In short, there would be a continuous, strong pressure of selection leading to the evolution of better organized, more efficient cells that could produce enzymatic systems at the direction of a single, or moderately replicated, coordinated pattern of nucleic acid molecules.

With the extension of the principle of natural selection to the preliving, chemical environment, Darwin's destruction of the grounds for teleological explanations in biology becomes complete. As Ernst Mayr[10] has pointed out, there are two ways of looking at the function of DNA. One is to trace from some part of the coded information transmitted in the DNA from each generation to the next the consequences, step by step, until one arrives at some end product. This is the *teleonomic* aspect, which is indeed all that Nagel discusses in his treatment of biological function and its apparent purposiveness. But the real question is why these causal chains leading from genes to end products should be appropriately fitted to the needs of life, why the chemical, physiological, and behavioral systems, like the structures, should be adaptive. This question Nagel ignores; but it is the heart of the issue. Its answer must lie in evolutionary biology, in the second way of looking at DNA. This is to study the changes that go on in the DNA itself from generation to generation, and to observe their hereditary consequences. The occurrence of such changes, or mutations, so the geneticist reports, are random, that is, are unrelated to their significance in the environment. After selection has taken place for eons, most of the changes in the DNA inevitably are destructive and harmful; yet occasionally a mutation may occur that aids an organism in leaving more offspring of its own kind than the

[10]Ernst Mayr, "Cause and Effect in Biology," *Science* 134 (1961): 1501–06.

average number of offspring produced by its competitors. Natural selection thus works, with almost infinite slowness but sureness, through differential reproduction. Adaptation to the existing environment is the result. Hence the history of the DNA explains its present state. Natural selection, before and since the advent of life, produces a semblance of purposiveness in the causal chains, though there has been only a blind, purposeless agency at work.

CAUSE AND EFFECT IN BIOLOGY

Henri Bergson, in *Creative Evolution,*[11] distinguished between three kinds of "causes" in an instructive way. In one case, the effect which is produced varies both quantitatively and qualitatively with the cause; in a second case, the effect varies quantitatively with the cause but not qualitatively; in a third case, the cause is merely a releasing mechanism, and the effect varies neither quantitatively nor qualitatively with the magnitude of the agent. One may illustrate these three cases quite simply.

1. A number of pierced billiard balls are strung on wires so as to be perfectly balanced and nearly frictionless. Then variable numbers of balls are placed in contact at the center of the wire, and impact is made by varying numbers of balls set in motion. It is readily observed that the number of balls which move from the center toward the far end of the wire is always equal to the number set in motion. In biology, this causal relation is represented by a chain reaction in which each step is both necessary and sufficient to set off the next step:

$$A \longrightarrow B \longrightarrow C \longrightarrow D \longrightarrow E$$

as, for example, in certain biochemical syntheses or in an unbranching chain of neurons. Such examples are rare in biology.

2. A rheostat is placed in the electric circuit leading to a lantern slide projector. By varying the rheostat, it is demon-

[11]Henri Bergson, *Creative Evolution*, trans. by Arthur Mitchell (New York: Modern Library, 1911).

strable that the image projected varies quantitatively with the amount of illumination, as regulated by the electric current; but the character of the image depends only on the particular slide which is interposed in the beam of light. In this type of causal relation, the cause is necessary, but not sufficient, to evoke the effect. Biology affords many cases of this kind, especially where branching of a chain-reaction occurs:

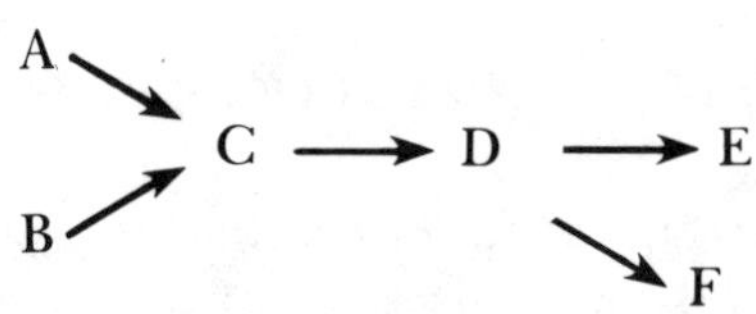

E may vary quantitatively with A or B, but the qualitative nature of the final response may differ greatly, especially when a transformation of energy is involved. Thus, to return to an example used earlier, the energy of contraction of a muscle is different in kind, being motion, from the chemical energy in the preceding terms.

3. The trigger of a pistol is pressed. Whether the pistol fires or not will depend upon whether it is loaded and whether the safety catch is open. The effect depends, both as to quantity and quality, upon the loading of the system. It has nothing to do with the energy with which the trigger is pressed, provided that it exceeds a minimum. The response is all or none. Probably the great majority of causal chains in biology are like this—at least the most familiar ones. All stimulus and response mechanisms where a change in the environment produces a response in the organism are initially of this sort, as are all reflexes such as a blink of the eyelids following a flash of light, a jerk of the hand when it touches a hot object, a flow of saliva when food is placed in the mouth.

Bergson goes on to say that only in the first case does cause really *explain* effect. Inasmuch as causal relations of the first type are far less numerous in biology than in the physical sciences, and far less numerous than the cases of types 2 and 3, we are presented immediately with one of the principal distinctions between the biological and the physical sciences. The physicist commonly attempts to set aside the role of the releasing mechanism or trigger and to press forward to an explanation in terms of causes of type 1. The biologist is often unable to get beyond cases of types 2 and 3 and, because of the complexity of the causal chains he faces, is basically interested in the release type of causation.

Braithwaite,[12] following Hume's thesis, which was adopted in succession by Ernst Mach, Karl Pearson, and many later philosophers of science, regards the idea of "necessary connexion" (that is, cause) to depend only on "constant conjunction." This position raises the question of the distinction between cause and correlation. Let us assume a causal chain, such as frequently occurs in biochemical and biological systems, of the following sort:

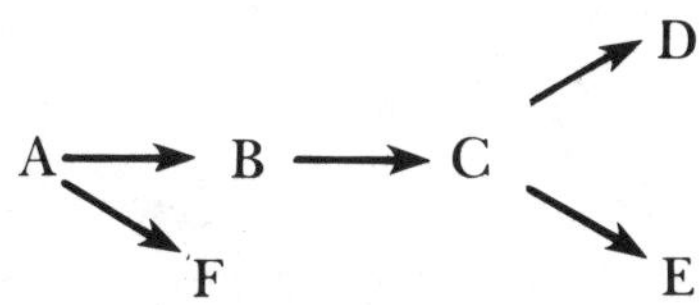

If F regularly follows A, and D and E also regularly follow A (but after steps through B and C), the criterion of constant conjunction is readily seen to be entirely fallacious. F will always be in constant conjunction with D and E and in addition will always precede D and E if the step A → B takes as much time as the step A → F. Yet what actually exists here is complete correlation between F on one side and D and E on the other and no direct causal relation between them. It may be pointed out that this situation is precisely what keeps going the still heated argument as to whether heavy cigarette smoking produces an increased liability to lung cancer. And, note well, that *insofar as prediction is concerned, it does not matter.* Complete correlation with an event, that is to say, "constant conjunction," is entirely as adequate a basis for prediction as direct causal connection. Mayr[13] has pointed out that Bunge, in his book *Causality,*[14] still maintains that the "touchstone of the goodness

[12] R.B. Braithwaite, *Scientific Explanation* (Cambridge: Cambridge University Press, 1955).

[13] Mayr, op. cit.

[14] M. Bunge, *Causality* (Cambridge, Mass.: Harvard University Press, 1959).

of a causal explanation [is] its predictive value," or in Bunge's own words: "A theory can predict to the extent to which it can describe and explain." Mayr's cutting remark on this is, "It is evident that Bunge is a physicist; no biologist would have made such a statement." The position assumed by Bunge is wrong on two counts. In the first place, I must agree with Mayr and with Scriven[15] that "one of the most important contributions to philosophy made by the evolutionary theory is that it has demonstrated the independence of explanation and prediction." As will be shown hereafter, the evolutionary theory very clearly explains, in terms of mutation and natural selection, isolation and adaptation, what it can in no wise predict. But there is a second fallacy in Bunge's position. To return to the example used above, where complete conjunction exists but no direct causal relationship, as between F and D (or E), let us suppose that the nature of A is entirely unknown. The observer would then be in a position, from the occurence of F, to predict D (or E), but he would lack any explanation for the seemingly causal relation of F to D (or E). The existence of A as a causal link effecting both F and D, in place of the theory that F causes D, can be demonstrated only if the step A → F can be blocked independently of the step A → B. To a biochemical geneticist this model is among the commonest sorts of branched system. The real steps in a direct causal chain are actually discovered in experiments by observing the action of mutations that block specific steps.

We may find it worthwhile to examine this example further, in terms of the three types of causes discussed previously. If A represents a compound that is convertible into either B or F, several possibilities exist. Both B and F may be produced. The two reactions will then compete with each other, in which case the quantities of B and F produced are less likely to depend on the quantity of A than on other factors in the total situation. Qualitatively, B and F may be similar to or different from A. A second possibility is that the production of F may have priority over production of B, or the converse. Then F, let us say, is produced until a sufficiency is formed, whereupon the step A → F is shut down and production of A → B is initiated. In this case the quantity of F is fixed and independent of the quantity of A,

[15] M. Scriven, "Explanation and Prediction in Evolutionary Theory," *Science* 130 (1959): 477–82.

but the quantity of B will depend upon the excess of the quantity of A above a certain threshold (determined by the amount needed to form F). This is an instance of regulation of a biological process which will be explained shortly. A third possibility is that of reciprocal regulation. In this relationship the steps A → F and A → B are alternately turned on and off. In such a case both F and B may vary in quantity directly with the quantity of A. It would seem to follow that one and the same causal chain or system can behave differently in respect to quantitative variation of "cause" and "effect." If so, then this aspect of causality is not essential, in spite of the fact that it is the classic type of causality sought by the physicist and long made the basis of mechanical determinism.

A word should be said about the developments in biology that in recent years have clarified the nature of the regulatory or reciprocal regulatory processes referred to in the preceding paragraph. These regulatory actions, in biochemistry or neurology, are now seen to result, at least most frequently, from positive and negative feedbacks in the causal system. The system is thereby converted from a chainlike array of steps to a circular process. For example, systems are known in which the end product inhibits the formation of the enzyme catalyzing one of the antecedent steps in the chain.

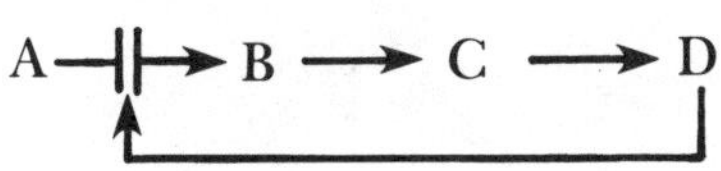

This is an example of so-called negative feedback. It regulates the amount of end product that is formed and prevents too great an accumulation of it.

On the other hand, in genetics more and more examples are being found in which the combination of a substrate molecule with an enzyme molecule *induces* the formation of additional supplies of the enzyme which at first is present in very small amounts. For example:

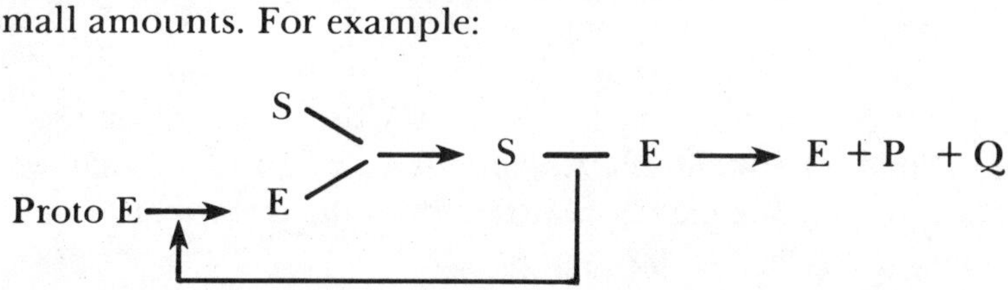

[where S = substrate, E = enzyme, and P and Q = products.]

This type of regulatory process insures that cells will not produce large amounts of particular enzymes in the absence of use for them. The phenomenon is a positive feedback. Such feedbacks probably lie at the basis not only of much physiological adaptation to changing conditions, but also of the developmental changes occuring in cells as the cells differentiate and become specialized. One of the major problems of current biology is to discern the nature of the forces or factors that bring genes into play at appropriate times during life and in different tissues. These releasing mechanisms are now thought to belong to the class of positive feedback mechanisms just described.

The "purposive" or teleological semblance of these systems that so marvelously control the development of an embryo into the appropriate type of adult plant or animal is thus purely physicochemical. Its regulatory aspect is that of feedbacks, acting analogously to a thermostat regulating a heating or cooling system or to the governor of a steam engine. Such are the causal systems whereby the DNA exerts its hereditary control over metabolism and development. They are appropriately called teleonomic to distinguish them from the evolutionary aspect of teleology which, as Mayr puts it, "deals more broadly with the over-all harmony of the organic world."[16]

INDETERMINACY IN BIOLOGY

To this point my argument has fully supported the view that the laws of physical science which are demonstrable in nonliving systems might be expected ultimately to explain all the laws of living systems. The distinctions made with respect to causality show up weaknesses in the conventional analysis based largely on physical systems rather than imply any basic difference in kind of causality. Yet there are good grounds for maintaining that the answer to our major question should be negative rather than affirmative.

Nagel has pointed out that the reduction of one science to another can be conceived only "as the deduction of one set of empirically confirmable *statements* from another such set"; that is to say, one may be able to reduce the laws of one science

[16]Mayr, op. cit., p. 1504.

to those of another but one cannot reduce one science to another by deducing *properties* from other *properties.*

> The deduction of the "properties" studied by one science from the "properties" studied by another may be impossible if the latter science postulates these properties in terms of one theory, but the reduction may be quite feasible if a different set of theoretical postulates is adopted.[17]

This seems very true, and yet I believe another aspect of the relation between the sciences is even more important. I refer to the statistical nature of scientific laws, the dependence of law upon probability rather than upon determinate certainty.

Bunge has discussed some aspects of this matter.[18] He says, "The realization of the statistical nature of all macroscopic laws, and of the objective nature of chance, has not led us to believe in uncaused events: it has only shown that not all relations are causal bonds." After defining what he means by "causal," he says: "There is always, as a matter of fact, some amount of spontaneity, mutual action, and randomness... *there are no purely causal laws.*" He proceeds to decry the equation "Causality = Lawfulness" and also the equation "Causality = Determinism." In the case of the first of these equations, it is clear that the sciences now abound with purely statistical as well as with classic causal laws. In respect to the second equation, however, the assertion that causality does not equal determinism seems to rest merely upon an arbitrary redefinition of determinism, such that even chance falls within the scope of "noncausal determiners."

Purely statistical laws are based on "random behavior," whether of the movements of molecules of a gas, the fertilization of a given population of eggs by a given population of sperms, or the number of "reds" and "blacks" that turn up in a simplified roulette game where only those two possibilities may occur. Randomness of behavior does not mean that specified conditions and measurable forces do not enter the situation—only that they are uniform for all the events. Thus, in an ideal Pascal

[17]Nagel, op. cit., Chap. 11, "The Reduction of Theories," p. 365.

[18]M. Bunge, "Causality, Chance, and Law," *American Scientist* 49 (1961): 432–88.

apparatus it is assumed that all balls are identical in mass and perfectly symmetrical in distribution of mass and in shape, that the apparatus is oriented true to the gravitational field, that all the balls fall from the same initial position with the same force, etc. It is, however, not necessary to make all these assumptions, since Poincaré demonstrated that in the simplified roulette game mentioned previously, "reds" and "blacks" would turn up with equal frequency no matter what the initial positions and velocities of the marble might be. The fundamental aspect of the situation is that there are always, in any real system, unmeasured random forces and unspecified random conditions that result in ordered distributions characterized by equal "choice" of equally probably alternatives. A perfectly flipped coin will fall "heads" and "tails" equally often, a perfectly balanced die will land with the 5-spot uppermost one-sixth of the time, and so on.

Statisical behavior of this kind exists at all levels of organization, from elementary particles to galaxies. It clearly operates at all biological levels, from the molecule to the organism, from the cell to the population, community, and ecosystem. The occurrence of osmosis in physiological processes depends upon the random distribution by diffusion of the molecules in a solution on both sides of a membrane, provided it is permeable to them. The Mendelian ratios depend upon the equal probability of an egg's fertilization by any one of two or more kinds of sperms when these kinds exist in equal numbers. The Hardy-Weinberg law, basis of all our modern evolutionary statistical predictions, is based on the random mating, with respect to certain specified characteristics, of individuals in a population containing different genotypes. Random behavior may be exhibited by the individuals composing a group that as a whole is undergoing an ordered shift, as in the Brownian movement of colloidal particles in a gas or liquid that as a whole is drifting in a current. Or conversely, a population as a whole may exhibit random behavior, as in the exploration of their local environment by a colony of numerous bees, even though the movements of the individual bees are very highly directed and determined.

It seems that statistical laws express an aspect of nature that depends upon randomness of behavior of the individual units whether these be atoms, molecules, cells, organisms, or higher entities in the levels of organization. What is more, the randomness of behavior of the units involved at one level does not necessarily depend upon the randomness of units at lower levels of organization. The statistical law does not depend on the relations between levels of organization. The probability that a particular face of a die will land uppermost is one-sixth simply because the die has six faces and not some other number; the probability that the egg will be fertilized by a Y-bearing sperm and give rise to a male, instead of being fertilized by an X-bearing sperm and so giving rise to a female, is one-half because the two sorts of sperms are present in equal numbers and are equally motile. If they are not present in exactly equal numbers or are not equally motile, then the probability will be different.

Such laws of science as these, arising directly from the nature of chance and the mathematical expression of probabilities, do not seem to be reducible to laws at a lower level of physical organization, because they describe the random behavior of entities at one particular level of organization. This is true both intraordinally (for example, within the biological sciences but at different levels of organization), and also interordinally (for example, between biology and chemistry, biology and physics, or chemistry and physics). A particularly clear example is afforded by the relation between the Mendelian laws and the Hardy-Weinberg law. Both express a statistical randomness in the transmission of genes from one generation to the next. The same biological processes are involved in both cases. But the Mendelian ratios arise from the randomness of fertilization of eggs provided by a particular female by means of sperms supplied by a particular male. The randomness here is at the level of the *cell*. On the other hand, the Hardy-Weinberg gene frequency equilibrium depends upon the randomness of mating between male and female individuals of various genotypes present in an interbreeding population. It depends upon the relative frequencies of these genotypes in the population, although the randomness of fertilization is implicitly assumed. One cannot derive the Mendelian ratios from the Hardy-Weinberg equilibrium, or the reverse, because the fundamental randomness each law describes is randomness at a different level from that in the other. The *random* behavior of sperms is simply not the same as the *random* behavior of mating multicellular individuals, even though sperms and mating obviously have a great deal to do with each other.

If this line of reasoning is correct, we may conclude that the statistical laws of one level of organization are not reducible to the statistical laws of another. By reduction directly to the mathematical laws of probability, any reduction to other levels of organization of matter and energy is obviated. In this sense we arrive at a negative answer to our main question: Can the laws of physical science which are demonstrable in nonliving systems be expected ultimately to explain *all* the laws of living systems? The answer is no—the statistical laws that describe random behavior are irreducible.

The second reason for doubting the reducibility of all biological laws to those of physics and chemistry is in fact a simple one, analogous to, but by no means identical with, the principle of indeterminacy in physics. It derives from the nature of the evolutionary process which affects all life.

The principle of indeterminacy in physics relates to single quanta of energy and single elementary particles. There is no reason to try to introduce this principle into biology, as some thinkers—I believe foolishly—have tried to do, in order to explain will or consciousness. Whatever material events lie at the basis of mental phenomena, they seem surely to depend upon a multimolecular basis. Like the events of chemistry, they involve the statistical behavior of millions or billions of molecules. But there is nevertheless a significant analogy between indeterminacy in physics and that in biology. The indeterminacy in physics arises because a single quantum or elementary particle cannot behave statistically at a given instant. It is unique, and the very operations performed to determine its position in space and its energy unavoidably alter the one or the other. In evolutionary biology, where one is concerned with past events—the mutations and the selections—that have led to the existence of a certain contemporary pattern in the DNA of a particular species, one is likewise dealing with uniqueness, the particularity of history.

Indeterminacy in biology is an outcome of the unpredictability of unique events. Predictability rests upon the statistical behavior, according to the laws of probability, of sufficiently large numbers of individuals. We may illustrate this principle very appropriately from the field of genetics, the branch of biology that is most advanced theoretically and most mathematical in its treatment of phenomena. In a particular kind of mating—let us say, one between a purebred spotted dog (such as a Dalmatian) and a purebred nonspotted dog—experiment and observation tell us to expect only offspring that are spotted. This results from the fact that all gametes from the one parent carry a gene for spotting, all gametes from the other carry a gene for nonspotting, and every fertilized egg will therefore carry a gene of each sort and be heterozygous. This rule predicts that the progeny will all be spotted rather than nonspotted. But predictability here, given such initial observations, is virtually perfect. Now let us cross the heterozygous spotted dogs together. Each dog produces half of its gametes carrying the one gene, the other half the alternative gene, or allele, as such a gene is called. Random fertilization between two kinds of sperms and two kinds of eggs, existing in equal numbers, leads one to expect one-fourth of the progeny to be pure spotted, half to be heterozygous (and spotted), and one-fourth to be pure nonspotted. If one makes a sufficiently large number of matings of this type, and produces litters totaling a hundred dogs or more, it is evident that this prediction is realized. Nevertheless, the prediction is a statistical one. Any single litter may deviate rather widely from the predicted proportions. The chance that a pup in such a litter will be spotted rather than nonspotted is 3:1, but the individual, particular event that produces a particular genotype in a particular dog cannot be predicted except as a probability.

The dog has many pairs of chromosomes, and the maternally and paternally derived members of each pair assort independently of those of every other pair in the special cell divisions leading up to the formation of mature reproductive cells. Thus, if we symbolize by A and A′ one pair of chromosomes, B and B′ a second pair of chromosomes, etc., a heterozygous individual that is AA′; BB′ will form four kinds of gametes: A;B, A;B′, A′;B, A′;B′. In mathematical terms, the number of kinds of gametes that can be produced is 2^n, where n is the number of pairs of elements assorting independently. The number of kinds of fertilized eggs that can be produced when two doubly heterozygous individuals mate is $(2^n)^2$. If n exceeds 20, these expressions become exceedingly large. For the dog, $2^n = 2^{39}$ = nearly 550 billion; for humans, $2^n = 2^{23} = 8,388,608$. The squares of these numbers are so large that it is evident that the potential numbers of kinds of offspring producible from a single sexual union are so vast that all the past, present, and future members of the species are not likely to approach it. The variety producible by recombination of the pairs of chromosomes is further increased by a process called crossing over, which permits pairs of genes that are linked, that is, are carried within the same pair of chromosomes, also to recombine. This is mathematically equivalent to saying that in the expression 2^n, *n* actually signifies the number of heterozygous pairs of genes which are present in each of the mates. This number may be very much larger than the number of pairs of chromosomes. Genetic variation is therefore almost infinite in any sexually reproducing population that is not strictly inbred. The genotype of each individual in such case becomes unique. Only by limiting attention to a relatively few pairs of genes can we achieve predictability.

The uniqueness of the individual is to be seen in biology not only at the molecular level of the DNA which constitutes the genotype. It also prevails at every other level. It is solely by abstracting from the infinitely varied individuals those elements and characteristics which they possess in common that the biologist establishes categories and species. As Mayr very well expressed it:

> [T]he individuality [that is] so characteristic of the organic world, where all individuals are unique; all stages in the life cycle are unique; all populations are unique; all species and higher categories are unique; all interindividual contacts are unique; all natural associations are unique; and all evolutionary events are unique.[19]

Uniqueness, as we have said, does not preclude prediction of a statistical sort, provided the populations and samples are sufficiently large. In fact, all scientific predictability is ultimately of this sort. But we now come to the final issue.

In the physical sciences, except at the level of quantum indeterminacy, one deals always with populations of enormous

[19]Mayr, op. cit., p. 1505.

numbers of molecules or other units, and the statistical laws predict with high accuracy. In astronomy, one deals with large numbers of orbital revolutions of a planet or satellite, or with a large and theoretically infinite number of observations of a radiation-emitting body. In biology, however, one deals with finite, and often very small, numbers of cases which, moreover, represent a history derived from previous finite and often very small populations of events. First of all, evolution is based upon the occurrence of mutations. These are exceedingly rare accidental alterations of the DNA, which thereafter are perpetuated. In a large population of organisms, every potential sort of mutation may occur, perhaps a number of times, in each generation. But in a small population a given type of mutation will occur, on the average, only once in many generations, and just when it will in fact occur remains unpredictable—this event being purely of the nature of a statistical accident. It follows that the fortunate mutations, in terms of the fitness and reproductive survival of a population subjected to new environmental conditions, may or may not, quite by accident, occur just at a time of selective crisis. The further fate of this population is thus determined by the random occurrence of an unpredictable event. We are forced to conclude that the evolutionary process is a causal chain of events the individual members of which are often—though not always—unpredictable and which in turn, delimit the possible situations existing in the next period of time. Like the flow of human history, which also contains a similar element of uniqueness of events, with the past limiting the character of the present and the future, there is limited usefulness of the evolutionary past for prediction. The probability mathematics of small numbers and unique events is simply not the same as the probability mathematics of large samples and large populations. That is the reason why the "explanations" of biology can in some respects not be reduced to the laws of physical science. Biological systems have a particular evolutionary history that conditions their present and their future.

I have discussed two of the four reasons Mayr has given for the existence of indeterminacy in biology, namely, the "randomness of an event with respect to the significance of the event," as in the case of mutations, and the "uniqueness of all entities at the higher levels of biological integration." I am not so impressed with the validity of his two other reasons, which are the "extreme complexity" of living systems, and the "emergence of new qualities at higher levels of integration." It seems to me that these reasons are akin to those that have previously been advanced and found wanting. If the arguments advanced in the discussion of the borderline between life and the nonliving are sound, then one might expect that eventually, and in theory, the laws of the physical sciences could be extended to explain systems of the utmost complexity. The emergence of new qualities is also not limited to living organisms. There is "emergence" in this sense when three kinds of elementary particles—neutrons, protons, and electrons—unite to produce an atom with new and unpredictable properties or when atoms of hydrogen and oxygen unite to produce a molecule with the unpredictable properties of water.

But the uniqueness of the particular event, embedded in the history and evolution of life, seems an unanswerable argument for the impossibility of explaining all aspects of life in terms of the laws of physical science which are demonstrable in nonliving systems. Finally, the randomness of behavior at distinct levels of organization leads to statistical laws based on probabilities that are applicable strictly to the entities that are exhibiting the random behavior. These statistical laws are explications as final at one level as at another. They relate to the nature of what we mean by "randomness," by "chance," and by "probability," rather than to the physical and chemical nature of material systems and measurable forces upon which our classical ideas of causation and determinism are founded. They find their unity in mathematics. They express the ultimate uncertainty of particular events in the real world where the laws of science deal only with the average outcomes.

4 The Centrality of Evolution in the Science of Life

Recent events in Arkansas (1981–83), Louisiana (1981–84), Texas (1984) and other states reveal that the battle to secure the right to teach the theory of evolution in the public school is far from won. Even the repeal in 1967 of the anti-evolution law in Tennessee marks only a turn of the tide. Following the famous Scopes trial of 1925 in Tennessee there ensued a long deadlock in this area of the freedom of thought. Most publishers of textbooks for the high schools and elementary schools, recognizing the possible danger to the adoption of their books in many localities if the books openly discussed the theory of evolution, and the even greater danger if they asserted its validity, decided to follow the course of least resistance. For a period of nearly forty years, while biological science itself was making vast strides and the modern science of genetics was becoming established soundly on an evolutionary basis, the biology textbooks of our schools referred timidly, usually in a final chapter, to the theory of the "development" of living forms. More often, the subject was omitted entirely, especially in the editions prepared for distribution in the Midwest and the South. Both the ancient and honorable *Biology* of Moon, Mann, and ..., which in its numerous revisions and changes of authorship

This essay was originally an address delivered to the annual convention of the National Association of Biology Teachers, meeting in Detroit in 1967. It was published in *The American Biology Teacher* Vol. 29, pp. 705–15 (1967), has been revised, and is republished by permission.

remained by far the most widely used biology textbook during this interval of forty years, as well as the newer and next most widely used book, *Exploring Biology,* by Ella Thea Smith, followed the rule of pecuniary discretion.

The Biological Sciences Curriculum Study (BSCS) was one of the first, and was the most diversified and widely adopted, of all the curriculum studies organized after the flight of Sputnik I in 1958 electrified the people of the United States and made us doubt widely that the American schools and system of public education were competing successfully in science and mathematics with those of the USSR. It was supported by grants from the National Science Foundation and was under the guidance of a board composed of university professors of biology, college teachers, secondary school teachers, and writers of broad experience. From its inception in 1959, the BSCS asserted the primacy of the theory of evolution in the teaching of biology. Unanimously the members of its steering committee, its writing teams, and all the BSCS workers agreed that, in Hermann Joseph Muller's phrase, "A hundred years without Darwin are enough." In selecting the themes which knit the study of life, otherwise a vast inchoate conglomeration, into comprehensible unity, the evolution of life was given first place. No fear of public opposition or rejection, no qualms about commercial success weighed in the balance against the deep concern felt by all of us that the best-tested and most coherent modern view of life and the nature of man should become the center and focus of our teaching. And this was done! In spite of some opposition in certain areas of this country to the adoption of the BSCS Blue, Green and Yellow Versions, all of which focus on the evolution of life and of man, substantial gains have been made in the freedom of teaching in Arizona, Texas, and Florida, among other states.

The earlier history of this conflict over the content of science textbooks has been covered admirably by Dorothy Nelkin.[1] More recently, after having received repeated rebuffs from

[1]Dorothy Nelkin, *Science Textbook Controversies and the Politics of Equal Time* (Cambridge, Mass. and London: MIT Press, 1977). See also John A. Moore, "Creationism in California," *Daedalus* 103 (1974): 173–89; "On Giving Equal Time to the Teaching of Evolution and Creation," *Perspectives in Biology and Medicine* 18 (1975): 405–41; and "Creationism," *The 1979*

the courts on constitutional grounds, the resurgent creationists have altered their plan of attack. They now claim that their interpretation of the origin of the earth and the life upon it is a scientific theory, and as such deserves equal treatment in the classroom. It seems not to matter that any such theory, however true it might be, is scientifically untestable, both unverifiable and unfalsifiable.

One notes that the self-styled scientific creationists adhere to only one version of the Divine Creation, that which is embodied in the book of Genesis in the Bible. No other versions of the Creation, equally revered in the sacred writings of religions other than Judaism, Christianity, and Mohammedanism, are considered by the scientific creationists to warrant equal treatment. Nor do the scientific creationists oblige by distinguishing which of the two quite variant versions of the Creation, described respectively in Genesis 1 together with the first four verses of Genesis 2, and in the remaining five verses of Genesis 2: 5–9, is to be supported. Nevertheless, bills to require that public school classrooms provide a "balanced treatment ... of scientific creationism and the theory of evolution whenever instruction centers on the origins of man, life, the earth, or the universe" (language quoted from one such bill pending in Louisiana) have been introduced and are still pending in no less than sixteen states of the Union during current legislative sessions. In Texas, one such bill has just been passed.

In 1981 also, a famous creationist suit came to determination before the California Superior Court in Sacramento. The plaintiffs, supported by the Creation Science Research Society of San Diego, were a student and his parents, who claimed that the teaching of evolution as a theory in the public school was "hostile to and contributes a repugnant coercion against the religious beliefs of the plaintiffs." It was also claimed that the California Science Framework's treatment of evolutionary theory was unconstitutional because "it constituted legislation pertaining to and encompassing the establishment of the religion of scientific humanism." The trial was marked by a rapid

AETS Yearbook: Science Education Society, (1978) 145–64. A newsletter, *Scientific Integrity*, published by the National Association of Biology Teachers, gives important coverage to recent developments in the creationist controversy.

retreat from their charges on the part of plaintiffs and creationist witnesses. Judge Irving Perluss dismissed the charges with an order of the court that there be disseminated to all publishers, institutions, school districts, schools, and persons regularly receiving a copy of the Science Framework a copy of the policy of the California Board of Education which states that editing be done prior to execution of a contract with a publisher to ensure "(1) that dogmatism be changed to conditional statements where speculation is offered for explanation for origins; and (2) that science emphasize 'how' and not 'ultimate cause' of origins."

Judge Perluss commended highly the words of William V. Mayer, the Director of the Biological Sciences Curriculum Study, who had appeared as a witness for the state in the trial, on the nature of science and scientific explanations, and the need for keeping one's scientific approach and one's theology separate.

Indeed, in all that I shall say hereafter one must keep in mind that a good scientific theory is simply one that explains satisfactorily the known facts, without being falsified by obvious conflicts with scientific evidence. It is always subject to repudiation when shown to be untenable by scientific observations or to modification to bring it into harmony with the facts, if that be possible. In the case of the theory of evolution, however, more than a century of such testing has made it a thousandfold stronger than in the time of Charles Darwin. One must, of course, make a sharp distinction between the theory of the evolution of life and the theory of natural selection as the chief agent of the changes. The latter, too, is supported by a vast array of evidence, but in spite of the "modern synthesis" of evolutionary views regarding the mechanisms of change, there are more doubters among scientists in respect to natural selection as chief agent than to the existence of the evolution of life and the origin of new species.

RELATION OF EVOLUTION TO ALL OTHER UNIFYING THEMES OF BIOLOGY

All the other unifying themes of biology are subsidiary to the theme of evolution—the change of living things throughout time. The old view of life, beyond the life and death of the individual and the limits of a single generation, was essentially a static view. The evolutionary viewpoint is dynamic. It sees living species in an unfolding panorama succeeding one another, fresh adaptations to the environment emerging on top of older adaptations and modifying or replacing them. It sees life emerging from the primordial chemistry of nonliving, nonreproducing compounds present in the warm seas of two or three billion years ago. It sees increasing complexity and the conquest of new environments, the land and the air. It sees the passage of the titans of old, those giant trees which formed our coal and which no longer exist upon earth, those dinosaurs and sea monsters that splashed in marine waters and in swamps, or fought their savage battles on the land. It sees the advent of the first amphibians to emerge on land, the first mammals to skulk in the undergrowth out of sight of predatory dinosaurs, the first primates that evolved an arboreal existence, and in the fullness of time the first ape-men and the first men, but little above the apes. It sees the dawn of intelligence in animate behavior, that is, an increased capacity to learn and to alter consciously the habits and ways of life; and it sees the beginnings of social order in tribe and family among the insects, where it is based on inherited instincts, and among the mammals, where it is based on learning, imitation, cooperation, and love. The grandeur of this view of life reaches its acme in the dawn of conscience and comprehension, in the achievement by man of an ultimate view of his place in terrestrial nature and the universe.

Most of what Thomas Henry Huxley and Charles Darwin thought of as "missing links" have been discovered in the ensuing century, along with a great many that they never imagined. The kinship of men of all races is clearly only a part, a very small part, of the total kinship of all living species of animals and plants and microorganisms. Man is a part of the whole, not an intruder. His evolutionary kinship validates the deepest insight of Saint Francis and of Buddha about the brotherhood of living beings. The split of a single homogeneous population into races adapted to different environments is a normal feature of the evolutionary process. Races may become extinct. Or they may continue to diverge genetically until they can no longer interbreed, until they become, in other words, fully distinct species. Or alternatively, races may reunite and merge again in a single interbreeding population, the diversity of each contributing to the diversity of the individuals in the combined amalgam. Man's future evolution, owing to his great present mobility, is clearly following this third pattern of evolution. The personal and social turbulence of the phase of intermixture is a passing aspect of human history.

The diversity of living forms together with the unity of the patterns into which they may be grouped on the basis of those characteristics they share is a unifying theme that is the root and substance of all scientific classification of living species. Without the operation of evolutionary change to explain it, the existence of such diversity and such unity would remain a mystery. In the framework of evolutionary theory it is on the contrary only to be expected: the unity of patterns results from the action of heredity, those species that share common features being descended from common ancestors. The most similar species have diverged most recently; those sharing only the most general and most universal features have diverged most anciently. The diversity arises, according to modern thought, from mutations of genes and modifications of chromosomes that occur accidentally and demonstrably in all species, after which they are subjected to the action of natural selection. Natural selection is a term given to differential reproduction, since some mutations and chromosome changes will make their possessors less viable and less fertile than nonpossessors. Thus, as Darwin saw and stated, hereditary variations when acted upon by natural selection will produce in time the elimination of the unfit and the preservation of favored types. The critical action in this process is exerted by the nature of the environment, which is the selective agent. In different environments, different types will be favored. Hence arise races, and ultimately new species. If the entire environment was uniform and unchanging, then, even though life might be present, it would presumably quickly have reached a standstill. It would never have led to a diversity of adaptations. The horseshoe crab *Limulus* and the brachiopod *Lingula* have scarcely evolved at all in their stable marine niches in the past 350 million years, while vast changes were taking place elsewhere. Man is the product of those lines that were subjected to tremendous environmental pressures, both physical and from other living things.

He belongs to the line that left the seas for the land and evolved the ability to breathe air and to lay a new kind of egg hatchable on land instead of in the water. It was a line that subsequently acquired hair and other means of regulating body temperature and thus of freeing one's level of activity from the fluctuations of surrounding temperature; that evolved differentiated teeth suitable for a varied diet, neither wholly vegetarian nor wholly carnivorous; that acquired a better arrangement of limbs beneath the body for support and for running; that evolved larger and larger brains which made possible the infinite variety of adjustments and controls of the environment whereby man now frees himself from most constraints of nature and swims, dives, runs, and flies beyond his ablest competitors.

Meanwhile, the human body retains relics of its past history and its past adaptations, beyond any reasonable accountability other than that of evolutionary change. Man's development is roundabout and possesses many stages that make sense only as heritages of the past. Vestigial structures abound, and others may form fully only to be reduced later to vestiges or to vanish altogether. The cave salamanders and fishes develop eyes that thereafter atrophy. The human embryo grows a tail and subsequently resorbs it. It is not quite true that every animal in its individual development climbs its own family tree, as Ernst Haeckel naïvely proposed. It is true that individual development is comprehensible only in the light of evolutionary history.

The genetic continuity of life, when fully understood in the context of modern biology, is likewise unintelligible without relationship to evolution. Sexual reproduction, so overwhelmingly the characteristic type of reproduction in plants and animals, and recently discovered to occur in bacteria and even, in a sense, in viruses, is meaningful only in the light of its relation to the existence of a diversity of hereditary types within each population of a species. Sexual reproduction, in fact, has the function of producing a variety of genotypes, and that variety is itself of value only in the context of the testing of many genotypes and the survival of few, through natural selection.

The science of population genetics, so strongly developed in the past forty years, is largely concerned with changes in the frequencies of competing genes within a population. The occurrence of mostly harmful mutations of genes and chromosomes would in the course of time lead to vast degeneration of the species if it were not in some way counteracted. The harmful genes are eliminated through the death or infertility of their possessors, the dominant ones at once, the recessive ones only after some generations. That is a description of the "survival of the fittest," the process of natural selection. Very rarely, a favorable mutation will occur, or one hitherto unfavorable may in some new or altered environment prove itself beneficial. More frequently, the selective process acts upon gene complexes. The existing gene combination AB may be favored, and the double mutant A′B′ may even be more favored; but the coincidence of the two independent mutations required to change from AB to A′B′ will be of excessive rarity. This is where sexual reproduction helps out, for the fertilization of an egg by a sperm, each derived from a different individual and line of ancestry, may provide shelter for the intermediate stages. First, if each parent contributes a gene of each sort, the offspring has a double dose of each gene, AB/AB, and is able to tolerate a mutation of A to A′ or B to B′, since one normal, effective gene of each sort is usually sufficient to provide normal biochemical function. Hence, after a time, a population may contain much heterozygosity; that is, there may be many AB/AB′ or AB/A′B individuals in it. Then by sexual reproduction, the next generation may contain A′B/AB′ and by recombination we may thereafter find that A′B′ eggs or sperms are produced, and A′B′/A′B′ individuals become possible. Since the effects of any gene are indeed dependent upon its associates as well as its environment, the consequences of the sexual processes in bringing together and recombining the genes in new genotypes really make rapid evolution practicable. Whenever the environment shifts, the concealed variability in the population is already available to provide thousands of genotypes to be tested out under new circumstances. Apart from the evolutionary process, sex makes no biological sense whatsoever.

Without natural selection, mutation would soon make a shambles of the organization of any species of living being. With natural selection, the elimination of harmful mutations keeps the species up to a successful level of operation and adaptation. With sexual reproduction, many of the otherwise harmful mutant genes can be kept in the population under cover until, in just the right combination or in just the right environment, they too become of value.

The complementarity of the organism and the environment is both evolutionary product and process. In a shifting and variable environment, the current adaptation is always being challenged and replaced in the population by some different, possibly better adaptation. The genetic processes operating in the past brought the current adaptation into existence, and at the same time provided the basis for future adaptations. In other words, no adaptation is once and for all, producing the best of organisms in the best of all possible worlds. The complementarity of organism and environment is dynamic. The environment, ever shifting, so varied in time and in place, is itself the selective agent. In a very real sense, the environment has thus created the organism, has shaped the species into what it is. The boundary between individual and its surrounding is in fact an illusion. The surrounding is a necessary part of the individual, in structure, in development, in reproduction, in all aspects of behavior.

The behavior of an organism is the totality of its responses to its surrounding. Some change of physical condition, some alteration produced by the presence of another living being, may as stimulus evoke a response such as a movement or a change of chemical activity. Most marvelous is the ability of the animal organism, and especially of man, to store up the impressions of past experience and to modify current responses on the basis of memory. All of these abilities to respond depend upon the organism's inherited capacities, which, as has already been stated, cannot conceivably be a static hereditary character of each species, uniform in every individual, but instead constitute a dynamic, evolving, genetic basis of life.

The complementarity of structure and function must likewise be interpreted in terms of evolution. Structure in the living organism is unintelligible without a knowledge of the function of each and every part; function on its side is unintelligible without a basis in structure, even at the level of the chemical syntheses and the activities of the enzymes within the cell. The cell is an ordered system, not a plastic bag full of homogeneously distributed contents. Its mitochondria, its plastids, its golgi system, its ribosomes all represent, like its nucleus and chromosomes, ordered systems in which certain chemical changes can take place at regulated rates without undue interference from other reacting systems within the cell. Similarly, the differentiation of cells and their assemblage into tissues and organs provide ordered systems at higher levels of organization. Yet all of these intricate aspects of structure and function are to be regarded as adaptations of the organism, adaptations that provide capacities to react and behave appropriately in the usual range of environments. Adaptations, furthermore, are products of natural selection acting over innumerable generations on the basic supply of mutations arising within a population, mutations themselves far less often adaptive than degrading and harmful both to structure and function. Mutations have been recognized by geneticists in the past largely as alterations of structure, but the present era is one of active exploration of mutations at the biochemical level. All changes of structure in fact flow from changes in the control of the genes over the synthesis of polypeptides (which are constituents of proteins). The polypeptides and proteins comprise on the one hand the enzymes that regulate all the biochemical machinery of living organisms and on the other hand the structural proteins, such as hemoglobin or myosin or collagen, that constitute the basis of all the living parts of every cell. Thus the mutations alter both form and function, whereas natural selection determines which mutations will persist through time in populations and species as the basis of all the adaptations that enable them to survive.

Since both form and function are repeatedly altered by genetic mutation, one could not even conceive of changeless species unless every mutation that occurred was eventually eliminated and no mutations under any circumstances were ever advantageous, or unless the hereditary types that exist within a species as consequences of mutation and selection never became isolated and differentiated into races and species. The evidence of the fossils and of the geographic distribution of races and species is entirely against such suppositions. It is quite clear that many times in the past a single species has become so transformed in the course of time that its descendants must rank as new and different species. It is also quite evident that geographic races do in fact arise within species, as indeed in our own species. While such races may become extinct or may reunite, very often they continue to differentiate until they become genetically isolated and forever lose their power to interbreed and merge. The old species has split into two or more new species when this has happened.

A classic example of this splitting of species into a greater number of descendant species is the astonishing variety of the finches on the Galapagos Islands in the Pacific Ocean, a situation first noticed and studied by Charles Darwin. It propelled his thinking inevitably in the direction of evolution. Each of these islands, depending on its size, is inhabited by three up to ten species of "Darwin's finches." Some have coarse beaks adapted for crushing seeds. Some have moderate beaks better adapted for eating insects. Some even have very slender bills with which they probe the hollows of cactus plants for insects. Each species is beautifully adapted to live in its own ecological niche where, on its island, it meets little competition from any other species of finch. But on different islands there are different species of finches than elsewhere. Now the finches cannot have lived for many centuries on the Galapagos Islands, for like the islands of Hawaii, these are volcanic in origin and recent in formation. Yet none of these species exists on the closest mainland, that of South America, or anywhere else, for that matter. It is therefore most likely that all the species of the Galapagos have descended from a few finches that reached the islands from South America, and finding habitats empty of competing types of birds, they evolved rather quickly into several species that spread from one island to the rest. The geographic isolation of the several islands reinforces the genetic isolation that prevents the newly formed species from intercrossing and thereby losing the refinement of the adaptations that have made them so great a marvel to all naturalists.

Finally, there is the biological theme of regulation and homeostasis. Claude Bernard's great insight, a century ago, led him to conceive that living organisms maintain in a matchless way the constancy of their internal fluids, the environment of the cells. Half a century later, Walter B. Cannon made many superb physiological studies that led him to redefine Bernard's "constancy of the internal environment" in terms of "homeostasis," a term signifying the regulatory capacity of the body to maintain a certain physical constancy by means of buffer actions and feedback circuits dependent on both neural and chemical responses. These most remarkable adaptive mechanisms exist as a part of the total organism, clearly under the control of the proteins and ultimately of the genes. Mutations of the genes affect the regulatory and homeostatic mechanisms no less than other parts of the organisms. Natural selection can be counted on to improve the regulatory and homeostatic mechanisms, which are in many cases related to changes of the environment, as in animals that live on land and breathe air in contrast to the animals that dwell in the waters.

No regulatory mechanisms are more fundamental in their nature than those which, perhaps as old as the dawn of life or of cellular beings, control the synthesis of proteins. It is, for example, now well known that many enzymes are not synthesized in cells that possess the capacity to make them, except when the substrates on which those enzymes can act are present. Again, the synthesis of certain enzymes is inhibited by the accumulation of the end product of the chemical steps in which the enzyme is involved. Thus the cell stops making quantities of such an enzyme as soon as the product becomes overabundant, whereas prompt and adequate utilization of the product keeps the machinery in full operation and the enzyme adequately replenished. In a multicellular organism the genes themselves are not active in every cell or at all times. They are turned on and off as differentiation proceeds. We now know of regulatory genes, of which there may be as many in the genome as the number of genes specifically concerned with the synthesis of proteins. The regulatory genes, in analyzed cases, themselves produce proteins that attach themselves to the "operator," which turns on a particular gene and which itself is an integral part of that gene. The operator is thus inhibited by the product of the regulatory gene and is kept turned off until, in the course of daily activity or of development, some substance is introduced that removes the inhibitor and thus turns the gene on. Regulation and homeostasis are thus ultimately gene-controlled. Modification of the genes by mutation, together with the natural selection of whatever forms of genes produce superior regulation, shape the nature of the regulatory system. Evolution here as elsewhere is an essential aspect of the theme.

Enough has been said to establish that evolution is the key to the understanding of life today. Neither the predominant feature of the panorama of life, its marvelous diverse adaptations, nor the central aspect of modern biological science, which is the study of the genes and the consequences of mutation in biochemical terms, can be comprehended without the assumption of the reality of the evolutionary process.

EVOLUTION—KEY TO UNDERSTANDING THE PAST

Neither can the nature of life in the past be understood apart from the evolutionary context. Abundant fossil links no longer missing now illumine the story of man's beginnings from the stage of his primate ancestry, through the stages of the wide-roving australophithecine man-apes of a few million years ago to the advent of the even more widespread ape-men now dignified by the cognomen of *Homo erectus* (once known as Pithecanthropus, Sinanthropus, Heidelberg man, etc.), and finally the succession of true *Homo sapiens* types (Neanderthal man, Cro-Magnon man, modern man). The succession of these types, now well dated by a variety of physical methods, parallels those cultural steps from tool-using to tool-making, fire-making, and burial of the dead that mark the upward course of intelligence from beast to man, with his foresight, his beliefs about supernatural powers, and his growing ability to mold his own environment. Can man shut his eyes to the implications of this now vast evidence of the nature and ways of life of his prehuman and early human ancestors?

EVOLUTION—KEY TO THE UNDERSTANDING OF THE PHENOMENON OF RACE

The races of man, of diverse geographic distribution, are a present phenomenon of profound social and historical significance. In addition to the five major ethnic groups generally recognized (Negroid, Caucasoid, Mongoloid, Australoid, and Amerindian), there are as many as fifty minor groups with anomalous or at least distinctive hereditary characteristics. Some of these ethnic groups appear to be on the way to extinction or absorption within a greater group, for example, the Pygmies of the Congo and the Pygmies of the Philippines, the Bushmen of South Africa, and the white and hairy Ainu of Northern Japan. Whatever the fate of such peoples may be, it can be stated as a major principle of human biology that all extant ethnic groups and races of man indubitably form a single species, for they all can interbreed with perfect fertility and without the slightest indication of any disharmony of developmental features in their hybrid offspring. On the contrary, there is considerable indication that the hybrids are remarkably good physical and mental specimens of humanity when measured against the average qualities of the ancestral types. There is only one exception to this principle known to me. *If* all primitive European people were Rh-negative,* as the present frequency of this blood type among Basques indicates may have been the case, and if all Rh-positive genes were introduced into the European populations as a result of successive waves of human migration from the east, where today the Mongoloid people are almost exclusively Rh-positive, then we might say that erythroblastosis fetalis, the hemolytic disease of the newborn, is a product of a certain kind of racial incompatibility. For the mother, of Rh-negative type, who becomes immunized by a leakage into her blood stream of red blood cells carrying the Rh-positive substance from her unborn child, would in that case represent an aboriginal European type (in this one respect), while the fetus would represent the hybrid Caucasoid-Mongoloid type. I hasten to add that the assumptions on which this interpretation might be made may be quite unjustified. We have really no assurance that the Rh-positive blood type was completely absent from the primitive European populations. We know only that it was probably far less common among them than in Europeans now, and that selection against the heterozygote favors a displacement toward a final population composed entirely of one type or the other.

These considerations lead us to a second major principle of human racial biology: the hereditary differences between different racial groups are rarely of an all-or-none sort. They

*Rh is chosen as a symbol for this blood type because the Rh substance was first detected in rhesus monkeys.

consist of different distributions and frequencies of the alleles (i.e., mutant forms of the same basic gene) in different populations. For examples we may again turn to the genes that determine the blood group systems, since these have been most widely and thoroughly studied in their ethnic distributions. The familiar ABO blood groups depend upon the presence or absence of three different major variants, or alleles, of the same basic gene. Two of these are common in every human group, namely, those that respectively produce the A substance or no substance (O). The rarest of the three alleles is that which produces the B substance and is present in persons of types B or AB. This allele, like the other two, occurs in every major race with the exception of the aboriginal Australoids and Amerindians. The frequency of the allele that produces the B substance (among all ABO alleles) varies from a high value of 25 percent among some Mongolian tribes to only 5 percent in Great Britain. It is now quite common among American Indians, but that is a secondary product of their racial intermixture with Europeans and Negroes. It is lacking in the Australian aborigines except in the extreme northeast and north of that continent, where it evidently was introduced by immigration from New Guinea and the Indonesian islands of persons of a different ethnic character. There are some five common Rh alleles, that is, forms of the basic Rh gene or gene complex. Only one of these, the R^o or *Dce* allele, possesses a markedly racial distribution. It is quite rare in most racial groups, for example, in Europeans and North American whites the frequency of this allele is only 2.8 percent; but it is the most common of all alleles among African Negroes among whom it ranges from 60 percent to 63 percent in frequency. The *r* or *dce* allele, which is the one involved in Rh-incompatibility between mother and fetus, has a frequency of 38 percent in whites and 19 to 24 percent in African Negroes, but is almost totally absent in the Chinese and Japanese populations. In summary, some alleles may be absent in some populations, but most alleles (except those that are very rare in all populations) are present in all populations in varying proportions.

For human characteristics that depend upon the interaction of many different genes insofar as their hereditary basis is concerned, as in the matter of stature or weight, intelligence or personality, it appears still more likely than in the case of single genes that the genetic differences between racial or ethnic groups are differences in the frequencies of particular alleles and rarely if ever are all-or-none in type. Take, for example, skin color. It is not certain how many different genes are involved in this trait. There are at least two, and perhaps as many as seven, acting cumulatively upon the permanent amount of pigment present in the skin irrespective of exposure to sunlight. If at each locus there are only two alleles, those respectively for presence or absence of deposition of melanins, the genotypes might range from a total of 0 to as many as 14 for melanin deposition. But the extreme genotypes are rare. Caucasoids might range from an occasional 0 representing extreme blondness to a maximum as high as 10, in some of the darker Mediterranean peoples such as Moors or Berbers, while the Negroes south of the Sahara Desert are by no means uniformly black skinned but might range in various tribes from 7 or less up to 14. Thus the racial distributions overlap. The ultimate conclusion to be reached is that the average difference between racial groups in respect to any hereditary characteristic is less meaningful than the actual difference between individuals, which may be—and often is—far greater within the limits of a single racial or ethnic group than any difference between racial averages. On the other hand, the possible presence within a single mulatto family of both very light and very dark children, and with other racial features such as type of hair segregating quite independently of the skin color, should warn us to be very tolerant in the judgment we place on the relationship between the phenotype of the individual and the common hereditary background.

EVOLUTION—KEY TO PREDICTION OF MAN'S FUTURE

What of the future? Prognostication of the future of man is quite impossible without a firm basis in our understanding of the evolutionary process, unless we take the quick and final step toward extinction through nuclear warfare. At the present day the genetic differences between human populations seem obviously to be disappearing in the course of increased migration and racial intermixture. Human mobility has reversed the geographic isolation that led in prehistoric times to the slow accumulation of different frequencies of alleles in different places. The races in time will disappear altogether, perhaps after only a few centuries, or at most millennia—a short period in the total evolution of man. The hereditary differences due to the presence of different alleles of each basic gene will persist in the common gene pool of the species and will be constantly replenished as new mutations occur. Selection will continue to operate upon the human genotype, although our increasing provision around ourselves of an artificial environment will substitute selection on the basis of the differential fertility of various genotypes for selection based largely on the basis of differential prereproductive mortality, which has prevailed in the past and still prevails for most of nature, plant and animal alike.

If man endures in spite of present dangers, he will soon be able to control the evolution of his own species as he already does that of his domesticated animals and his cultivated plants. The mechanisms are clear, the methods understood. Only certain techniques remain insufficiently developed, certain taboos constrain. Evolution within the limits of the species is fundamentally differential reproduction, for the more abundant reproduction of certain genotypes and the less rapid reproduction of others is what leads to the differences of gene frequency that in the past have characterized different geographic races.

The goal of medicine, to put it baldly, is the elimination of all differential mortality, whether resulting from nutrition and other environmental influences or resulting from differences in genetic endowment. The physician, by the traditional ethics of his profession, is committed to preserve life whenever possible. The more completely he is successful—and already in the more advanced countries of the world he has been almost completely successful during the all-important prereproductive years of life—the more surely will the bearers of once harmful genes live to reproduce and transmit those genes. One might say that this is quite all right and will produce no bad consequences, with two provisos: first, that the social cost does not become too great; and second, that the artificial environment thus constructed be guaranteed permanency. A sudden reversal of the ameliorating conditions might be disastrous for the possessors of all sorts of genetic defects. Thus, in the chaos after a nuclear war, survivors totally dependent upon regular supplies of insulin or other hormones, drugs, or devices would find that natural selection was restored in utter harshness. Even the nearsighted and astigmatic whose spectacles had been broken might be unable to survive for lack of adequate vision! As for the social cost, the doctors are persistently making more work for themselves, and it will therefore always require more doctors just to keep the population at the same level of well-being.

From the foregoing considerations we arrive at two conclusions. In the first place, even if we make no deliberate effort to control the future course of human evolution according to plan, the practices in which we are now engaged are surely and certainly changing the composition of the human gene pool. The migration and interbreeding of peoples increases the number of heterozygotes in whom recessive genes remain unexpressed and escape natural selection. Interbreeding, which promotes homozygosis and a relatively rapid elimination of harmful recessive genes, is yielding everywhere to panmixia, a breeding structure of the population that permits much increase in the absolute number of harmful genes before their frequencies become sufficiently high that the probability of homozygous offspring reaches the earlier level. As in the influence of the medical effect, this change does not itself produce any increase of overt harm. Only if, for any reason whatsoever, the mating system was to undergo reversal, shifting from panmixia back to inbreeding, would there be dire consequences, for in that event the number of homozygotes for harmful genes would immediately be greatly increased, in proportion to the degree of inbreeding.

In the second place, then, both from the effects of medical ethics and from the change in the breeding structure of human populations, it is clear that we cannot afford to turn the clock back. Our evolution is on a one-way street. The abilities we may lose or sacrifice in the course of adaptation to our brave new world can hardly be regained, except by artificial means. The air-breather may return to live in the ocean, and its limbs may be remodelled into paddles, but it has never been able to reevolve the useful gills that enable a fish to extract dissolved oxygen directly from the water. The animal that once adopted a predatory life to live at the expense of the green plants so abundant in its world lost the ability to synthesize all sugars and amino acids, and forever remains dependent on the plants for food. Man has been a fortunate species thus far in his evolution, for many of his own adaptations have increased his ability to live in widely varied environments and to utilize many different kinds of food. Above all, his intelligence makes many of his adaptations flexible and temporary rather than fixed and permanent. But evolution is strewn with the relics of extinct species that were not so successful, that evolved rapidly and successfully in a way that restricted their future way of life more and more, until eventually some change of environment wiped them out. The species that becomes restricted to one particular food plant, for example, is at the mercy of the survival of that food plant. The species that can live only in a cool or cold climate disappears with the end of the Ice Ages. Man should take warning, for his choices, whether unconscious or deliberate, will foreclose certain avenues of genetic evolution.

At the present time, mankind the world over is moving very rapidly toward conscious control of the reproductive process. The new steroid hormones and intrauterine plastic loops are but harbingers of greatly improved methods of contraception. Perhaps these are the only means available whereby man can avoid the imminent threat of overpopulation and its Malthusian consequences. Nevertheless, the benefits of introducing universal contraceptive methods are not unalloyed. We may further expect that the reproductive process itself will be transferred increasingly to the laboratory. Techniques are already developing that will permit the culture of ovarian and testicular tissue in the laboratory, a continuous production of mature eggs and sperms, the formation of embryos by fertilization "in vitro," and the implantation and culture to term of the embryos either by implanting them in a foster mother or growing them entirely within an artificial womb. There will result, in the long run, a total severance of the human sexual life from the reproductive process, which may be combined with some form of controlled mutation and of selection for absence of defect and presence of desired qualities. Removal of the sex life from all connection with reproduction may well threaten certain basic human institutions, especially the existence of the family as the fundamental unit of society.

Such is the speed of scientific progress that many of these possibilities will probably be realized before the close of the twentieth century. We have little time and very little foresight and wisdom to plan for such a future. On one matter, however, we can be positive. The ethics of a static human society cannot cope with an evolutionary situation. It must be replaced by an ethics that takes into account the evolving biological as well as cultural nature of man. Our growing wisdom must be based soundly on an evolutionary view of man's past, present, and future, and a knowledge of the ways in which the evolutionary process can be controlled.

THE GOALS OF HUMAN EVOLUTION

To what end? The choice of goals is not for the biologist alone, but for all men—yet the seer is indispensable, and without evolutionary insight there is surely no seer. Aristotle, who was a great biologist before he was a philosopher and whose biological insight sharpened and shaped his philosophy, conceived of every action as moving toward a Final Cause. He could not conceive how an embryo chick, from a formless substance within the egg, could develop into a male or female adult bird without the "Final Cause, the End or Object towards which a formative process advances, and *for the sake of which* it advances—the *logos,* the rational purpose" of its being.[2] Aristotle thus gave us teleology, which has plagued biology to this day with its imagined need for inner, directive forces that drive or draw all life to appointed ends.

Darwin freed us lastingly from teleology. Natural selection, he affirmed, will create all the adaptations of species by the preservation and differential reproduction of the favored types, that is, those with advantageous hereditary variations. Our

[2]A.L. Peck, introduction to Aristotle's *Generation of Animals* (Cambridge: Harvard University Press, 1943), p. xxxviii.

increasing knowledge of the participation of mutations of the genes and chromosomes in this process during the past half century has strengthened our belief in the creative aspect of natural selection. Natural selection does not create the mutations, but without them it would lack hereditary variations to act upon. Natural selection, moreover, acts on individuals or groups of related individuals, not on separate isolated genes, for it is the individuals that reproduce themselves. Consequently it is the entire genotype of an individual that forms a basis of selection, yet only indirectly, for actually selection must act on the phenotype of the individual, on its totality of manifested characteristics as they have resulted from the presence of a particular genotype governing development within the constraints of a particular environment. Selection acts through mortality on individuals at every step during this early development, as well as through both mortality and fertility during the reproductive period. What is thus selected is a constellation of genes that work harmoniously together and by degrees produce adaptations that fit the organism at every stage of life, immature or mature, for a particular environment in which that portion of the life is normally spent. It follows that every stage of development seems to anticipate the next. The mature fowl is indeed potentially implicit in the genotype of the egg. Nevertheless, Aristotle was wrong. The genes, given nutriment and appropriate conditions, produce a chicken; and the genes are materially present in the fertilized egg. There is no final cause that draws the egg to its foreordained goal, only the semblance of moving toward that goal. The semblance exists because of the appropriateness of each part and each process during development. That, we now understand, is the product of the evolutionary process, the age-long selection among all the genetic variations that have actually occurred, of just those that permitted the survival and favored the reproduction of their possessors. It has been proposed by Ernst Mayr[3] and generally adopted that we call this semblance of a teleologic development a *teleonomic process*—one in which the seeming goal is achieved because of the work of natural selection in the past instead of mystical processes and powers in the present.

[3]Ernst Mayr, "Cause and Effect in Biology," *Science* 134 (1961): 1501–06.

The ultimate concern of man, said one of the greatest of modern evolutionists, Theodosius Dobzhansky, is with man's search for the meaning of existence.

> Among the two million or more species now living on earth, man is the only one who experiences the ultimate concern. Man needs a faith, a hope, and a purpose to live by and to give meaning and dignity to his existence.... It is, furthermore, inconceivable for mankind to have meaning if the universe has none. Man is involved in mankind, mankind in life, life in the planet, earth, and earth in the universe.
>
> To satisfy man's hunger for meaning, not only man but the whole of nature, living and non-living, must be understood in their relatedness. For man, though he may be nature's spiritual vanguard and spearhead, is nevertheless only a small part of nature.... Life is very much older than man, and the universe is much older than life. This points to an indispensable condition which any synthesis must satisfy in order to be acceptable. It must envisage man, life, and the universe as changing rather than fixed, as parts of a single ongoing process rather than as three separate static realms. The central postulate of the synthesis must be that the universe and everything in it are evolving products of evolution.[4]

This is a biological philosophy of grandeur and exalted faith. The meaning of man's life may lie outside his biological experience, yet his ultimate concern cannot remain incompatible with scientific truth. Man's evolving nature gropes for assurance regarding the why of existence, and evolutionary theory sends a shaft of light to illumine his past and present, and to penetrate somewhat the mists of his future. As he stands with the power of the evolutionary process in his grasp, man sees in his own nature dissension, aggression, and combat arrayed against unity, cooperativeness, altruism, and love. Is there assurance of the future course?

Ages ago, when man first acquired some knowledge of the good or evil consequences of his actions upon his family and

[4]Theodosius Dobzhansky, *The Biology of Ultimate Concern* (New York: New American Library, 1967), p. 110.

his tribe, his dawning conscience quickened in him a sense of sin. Slowly thereafter, as the comprehension of human involvement has grown, some men have begun to call all other men their brothers. Like Saint Francis long ago and Albert Schweitzer in recent times, the biologist of the present day claims the sympathy and concern of man for all life, since all are fragments of the grand evolving web of beings. And now man himself has become the agent of the evolutionary process, the trustee of all life here on earth. To fill that role wisely he has much to learn, in spirit as deeply as in store of knowledge. Without a central place for evolution in our teaching and our learning, we cannot even begin to remold our world and our own natures in the direction of our goals. We will instead play madly and blindly with forces that may destroy us all.

5 The Genetic Basis of Human Racial Differences

Of all the biological problems that today face the world, the most severe are probably those of population, food, and race. These are by no means independent problems, since the geographic distribution of peoples of different races in more favored and less favored parts of the earth, in ancient centers of civilization, or in underdeveloped lands, tie the problems of racial differences to those of a sufficiency of food; and so far in human history the growth of population has only become a threat to subsistence whenever the food supply is inadequate for the growing population. Some day—not too far off—mere living space and the insufficiency of other biologically necessary resources, such as water, will perhaps become paramount. But today, although in certain parts of the world these last-mentioned factors are locally critical matters, they are not major problems. Social biology must occupy itself far more actively with the solution of the quandaries presented by population, food, and race in the immediate future.

Of these three major problems, those arising from race differences fall most clearly within the purview of a geneticist, and they form our present subject. The existence of races within the human species can scarcely be denied, even though some anthropologists and sociologists seem to have thought that by

---

This essay was first prepared as an address to Beta Beta Beta at its triennial convention in December 1959 and was published in *BIOS* 31(1960):1–20. It has been revised for the present volume and is reprinted by permission.

denying the existence of human races we might very simply do away with the problems they pose. The formation of races is a normal part of the evolutionary pattern in the formation of new species. Whenever populations belonging to some single species become geographically isolated from one another, so that individuals of the different populations thenceforth rarely or never mate and produce offspring, certain consequences are to be expected. Mutations will occur in genes and chromosomes quite independently in the two or more isolated populations. If certain kinds of mutations are very rare, for example, of the order of 10^{-8} per gamete (1 per 100 million), and if the populations are not so very big, say of the order of 10^5 (100 thousand) individuals per generation or fewer, then such mutations will not be expected to occur in a particular population more often than once in 500 generations. Consequently, the same mutations are not at all likely to be represented in different populations. Even though some mutant genes may be shared in different populations, the majority of the more rarely mutating ones will be different, and by the random nature of genetic change, and that alone, the populations will tend to diverge in type.

Natural selection will also enter the picture as a potent force leading to differentiation. It is most unlikely that in physical geography, climate, kinds of food, and ecological relationships between the species already living there the different areas in which the populations are isolated will be exactly similar. To the extent that the areas differ in the conditions of life, the collections of different mutations that arise in the isolated populations will be forthwith subjected to different selection pressures. Except in very small populations, of a few hundred breeding individuals or less, these selection pressures will be the dominating evolutionary forces. They will determine which alleles (alternative states of genes), at each and every genetic locus of the chromosomes, will persist in the population and which alternative alleles will be eliminated, which will be common, and which will be rare.

In the course of time—partly through the selection of random differences between the isolated populations and partly because whenever such adapted types renew contact with each other and begin to hybridize there is a breakdown of the adaptedness in their hybrid offspring and consequently a ten-

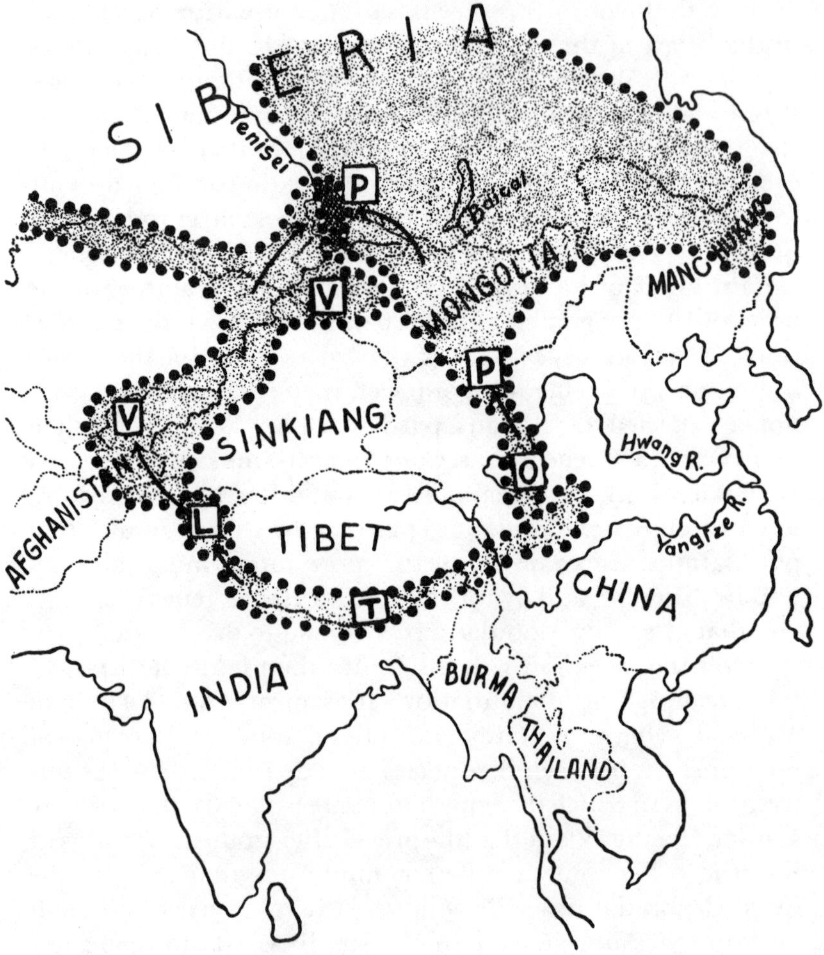

FIGURE 5.1. A ring of races in the warbler *Phylloscopus trochiloides*. The letters V, L, T, O, and P represent the several geographic races forming the ring. All of these interbreed with one another where they come into contact, except for V and P. In the shaded contact region between these two races, both of them inhabit the same area without interbreeding. (From Ticehurst).

dency for natural selection to preserve anything that will prevent or reduce intermating—there will arise some degree of mutual intersterility, or some aversion to mating that will tend to minimize the intermixture of the differently adapted types. Eventually, this process reaches a point where members of the two groups will no longer intermate and exchange genes. Their gene pools, we then say, have become fully isolated. Many examples of different stages in this process are known to zoologists, among the most dramatic of them being the rings of races occurring around some central, climatically unsuitable areas. In a typical ring of races, race A will interbreed with race B to some extent at the boundary between their territories, race B with race C, race C with race D, and so on. If race E or F completes the circuit and comes into contact with race A, it often turns out that individuals of races A and F will not interbreed, or if they do so, will not produce viable or fertile offspring. The transfer of genes between the races, or gene flow as geneticists call it, is then shut off at this point, although mutant genes present in race A might still slowly make their way around to face F by seeping into race B, then into race C, and so on, around the long side of the ring. Good examples of such rings are illustrated in species of birds that occupy territories around the central inhospitable area of Tibet (See Figure 5.1), or around the North Pole in a circumpolar orbit, or among salamanders in the hills and valleys surrounding the Great Valley of California (See Figure 5.2).

It will be apparent that if any link in such a ring, or chain, of races should break, that is, if some intermediate race were to become extinct, then the gene pools of the terminal groups of races would become completely isolated geographically, except in the territory where forms A and F meet and do not interbreed. No systematist would then hesitate to identify the two groups as two separate species. A ring of races thus graphically illustrates the gradual evolution of geographic races into distinct species. It enables us now to arrive at a fairly satisfactory definition of a species on a biological, or genetic, basis—though of course our definition will apply only to species that reproduce sexually and evolve in this geographic way. The species, under such restrictions, is that level of evolutionary differentiation at

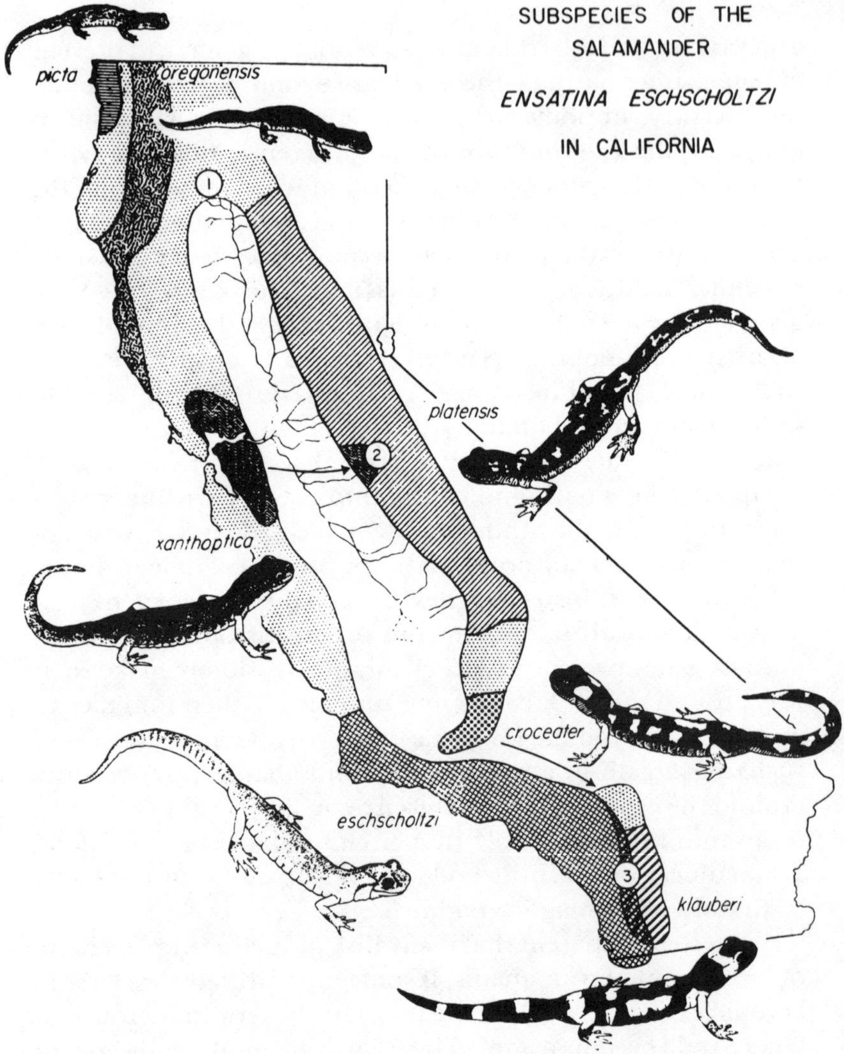

FIGURE 5.2. Races of the salamander *Ensatina eschscholtzii* in California. The differently shaded areas represent the habitats of the several races depicted, and the areas of overlap and intergradation between them are stippled. (1) An area where there is a smooth intergradation. (2) An area in which two races, living closely adjacent, hybridize infrequently. (3) An area where two races have reached the level of incipient species, since they inhabit the same territory but do not interbreed. (From Dobzhansky, after R. C. Stebbins).

which the gene pools of the populations in question can undergo little or no interchange, the level at which gene flow is impossible for genetic, and not merely geographic, reasons. The geographic race then, in contrast, represents a lower level of evolutionary differentiation, one at which the gene pools of the populations have become differently adapted to their local habitats, but at which gene flow between the populations is still possible whenever they come into contact.[1]

We may now apply these concepts to our own species, since *Homo sapiens* obviously fits our requirements of being sexually reproducing and genetically differentiated into separate geographic populations. Experience is ample to show that there is no incapability on the part of members of any human populations to interbreed and form fertile offspring. The hybrids, according to the most critical studies, as for example in Hawaii, exhibit no deficiencies for living in the present human environment—in fact, they seem to possess a certain degree of hybrid vigor and some superiority of physique and, some persons would say, beauty, in comparison with the stocks undergoing mixture. Consequently we can discard any idea that mankind is now divided into different species, even though this notion has been seriously put forward by a few modern anthropologists and geneticists (e.g., Carleton S. Coon in *The Origin of Races*).[2] On the other hand, evidence is abundant that there are many genetic differences between human populations that have been long isolated in different continents, and even within the limits of a single continent. How fine one wishes to divide up the species into ethnic groups denominated "races" is perhaps a matter of preference. I believe that most geneticists and anthropologists today are fairly well satisfied with a subdivision into four or five major geographic races: the Mongoloid, the Negroid, the Caucasoid, and the Australoid, to which is some-

[1]Fuller expositions of these concepts of geographic speciation may be found in such works as *Genetics of the Evolutionary Process*, by Theodosius Dobzhansky (New York and London: Columbia University Press, 1970); or *Populations, Species, and Evolution*, by Ernst Mayr (Cambridge, Mass.: Harvard University Press, 1970). Although newer facts and theories have been added, the fundamental concepts have stood the test of time. It has seemed to me better not to modify the original treatment more than necessary.

[2]Carleton S. Coon, *The Origin of Races* (New York: Alfred A. Knopf, 1962).

times added the Amerindian. There are of course many difficulties if one tries to assign certain relatively small groups, such as the Ainu, to one of these five races—but that is the way of evolution, not clean and neat and systematic according to the pigeon-holing proclivities of a human mind, but diverse and almost chaotically varied in a ramifying whole. There are many microgeographic races as well as the major ones, not only the Ainu, for example, but the pygmies of Africa and the Philippines, the Polynesians and Melanesians, and many others.

As we begin to examine the genetic nature of the major races, the salient feature is that differences are very rarely "all or none." Far more commonly, each race under comparison will possess each sort of allele known to occur. For example, at the locus which determines the ABO blood groups there are three principal alleles, I^A, I^B, and I^O. All three of these are to be found in each of the major races, with the exception of I^B, which is lacking in the American Indians and the Australian aborigines. M and N occur in all of the races. Of the commoner Rh blood group alleles, R^1, R^2, R^O, and r are present everywhere, except that r is almost entirely absent from the Melanesian, Australoid, and most Amerindian peoples, and R^O is rare among Europeans and some other peoples. The examples could be multiplied, but those already given suffice to show that the unique gene, which may be said to characterize a particular race or group of races, is the exception, the rule being that races differ mainly in possessing different frequencies of the same genes. This probably applies also to the genes responsible for the superficial characteristics of skin color, eye color, hair color and type, conformation of body or skull, and the like. In all such features, as Darwin noted, the races "graduate into one another." It is very probable that the genes on which such traits depend will, when better analysed, be found to have a graded distribution, too.

MUTATION

Let us turn to a consideration of the evolutionary forces that produce the racial differences. The first of these is the mutation process, for without it there could be no different alleles at any genetic locus—all individuals would be homozygous and all would be identical. Whether we speak of individual hereditary differences or of racial differences, these depend alike on the capacity of the genes and chromosomes to undergo mutations. During the past fifty years of study of induced mutations in experimental plants and animals, it has become clear that within any population of a species any sort of mutation known to occur at all in the species can probably occur, although some mutations may occur more rarely in certain races than in others, perhaps because of the presence of suppressor genes that prevent such mutations from manifesting themselves. Even in different species, if they are very closely related, the same kinds of mutations occur and can be located at corresponding sites of the chromosomes or can be shown to be alleles by their interaction in hybrids between the species. Thus we can say with some confidence that many of the same genes exist in different species if these are not too distantly related. In the human species, up to the present time, the only detection of actual mutation events relates to rather rare dominant types, such as kinky hair in a blond Norwegian girl, or the type of dwarfism known as achondroplasia, and the like. But very often the pedigrees are not well authenticated for more than a generation back, so that the origin of the dominant trait is uncertain. When a trait is recessive, like albinism, for example, its sudden appearance in the population in a child born to normally pigmented parents is not usually attributable to an immediate mutation, since the mutant gene has to be transmitted by both parents for the trait to appear in a child, and most likely it has then been transmitted through a good many generations of ancestry on both sides of the family. When a characteristic is sufficiently detrimental to lower the probability that its bearer will reach adulthood and reproduce, the mutant gene, whether dominant or recessive, will gradually be eliminated from the population, the dominant genes more rapidly, the recessives more slowly. This process must needs end in the elimination of all harmful genes in the course of time, were they not in some way replaced. Hence modern theory holds that the pressure of selection tending to eliminate harmful genes must be balanced by an equal and opposite mutation pressure tending to restore the harmful genes to the population. But since the pressure of selection becomes stronger as genes are more harmful, the equilibrium that results between mutation and selection reaches different levels for different genes. A mutant allele producing blue eye color or red hair or blood group *B* may become quite common in a population, since selection against them is mild if not negligible. But a mutant form producing six fingers and toes, or albinism, or colorblindness tends to be much rarer. Later another possible way in which harmful genes may be kept in a population will be mentioned, but enough has been said already to make a major point about racial traits. Since they are common, they cannot be very detrimental. There is a good chance that they actually confer a benefit on their possessors, at least within those original environments in which the races evolved.

A very good example of a mutant trait that must have originated not so very long ago, and that since its origin has manifested superiority in the sense that it has spread, is the occurrence of reddish, tawny hair in some of the tribes of Australian aborigines.[3] This hair has assuredly not been derived from wandering European sailors or gold hunters. In two tribes living in the heart of the Australian Desert, every individual has hair of this sort. In the regions round about, the percentage of tawny-haired natives is not quite 100 percent; farther away in every direction it is less; and so it diminishes in frequency until finally one reaches areas where the gene has not penetrated. A diagram with lines connecting places with equal frequencies of this trait in the several populations shows very strikingly the distribution of reddish hair, as well as its tendency to spread more in some directions than in others (see Figure 5.3). Why this trait should evidently succeed in displacing non-tawny hair in these tribes we cannot say, but through some remarkable property the bearers of this gene must be more successful in reproduction, must produce more successful offspring, than those who lack tawny hair. If this process continues for some generations into the future, we may actually see the entire Australoid race acquiring a new racial characteristic, that of reddish hair. Yet even if this were to happen, the new racial

[3]This unique and interesting investigation was reported by the American anthropologist Joseph B. Birdsell, "Some implications of the genetical concept of race in terms of spatial analysis," *Cold Spring Harbor Sympos. Quant. Biol.* 15(1950):259–314.

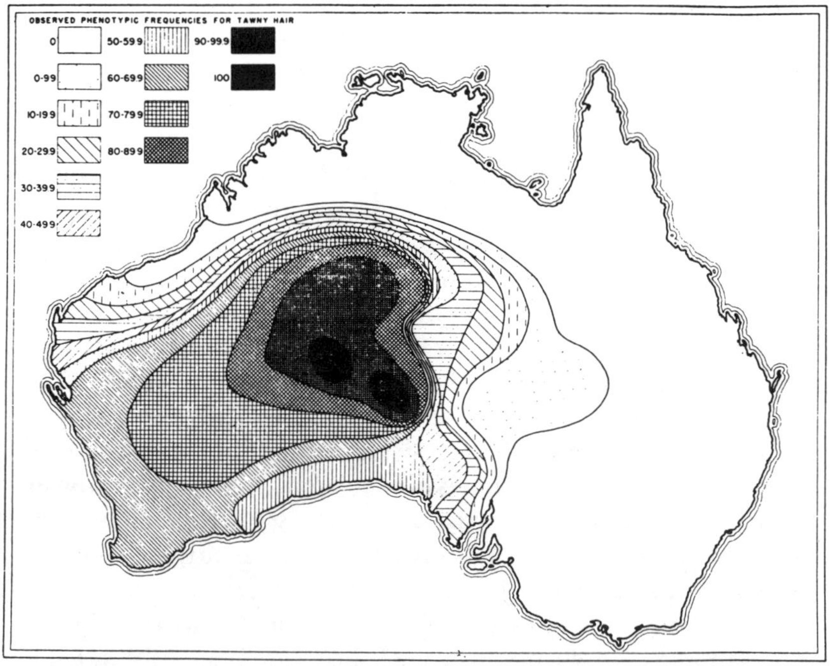

FIGURE 5.3. Distribution of tawny hair among the Australian aborigines, to illustrate clines radiating from the center of origin of a mutation. (From Birdsell).

characteristic would not be unique, since red and reddish hair also occur in peoples in other parts of the world.

SELECTION

It has been impossible to discuss the role of mutation in the origin of racial differences without already introducing the role of selection. Some further examination of different aspects of selection may now be made.

One fundamental feature of human races is their polymorphism, that is to say, the existence within each race of a variety of alternative types, resting on the existence at numerous genetic loci of two or more alleles that give rise to the alternative phenotypes. Different inherited shades of skin color among whites, different eye colors in many populations, different blood groups—these are but a few of many examples that might be chosen. Polymorphic traits are not the rare hereditary characteristics maintained in a population by mutation alone. They are relatively common traits that would seem to be either neutral or, under certain circumstances, positively advantageous. Sometimes an inherited trait that seems obviously detrimental is nevertheless so common in certain populations as to raise a real question. Either it would take an extraordinary rate of mutation to raise it to such a frequency, or it must, in spite of appearances, confer a positive advantage in some way.

The occurrence of sickle hemoglobin is already a classic example of this phenomenon. Sickle hemoglobin occurs chiefly among Negroes, although it is present around the Mediterranean Sea and in places in tropical Asia. Strictly speaking, it is neither dominant nor recessive. A double dose of the gene that produces sickle hemoglobin results in a severe, often fatal anemia, since the oxygen-carrying capacity of the abnormal hemoglobin is low. In a person with one dose of the sickle hemoglobin gene and one normal allele, sickling of the red blood cells may be observed when the oxygen tension is reduced, but there is no anemia. Now in certain African tribes the frequency of persons with sickle hemoglobin exceeds 45 percent, a fact making one wonder how a gene so harmful when homozygous can nevertheless be so common.

In the 1950s a young British physician, A.C. Allison, undertook to test a suggestion that had been made a number of times, namely, that the geographic distribution of sickle hemoglobin might be related in some way to the prevalence of malaria.[4] He tested Negro subjects possessing some sickle hemoglobin and others with only normal hemoglobin by inoculating them with the plasmodium of tertian malaria (*Plasmodium falciparum*) or by permitting them to be bitten by infected mosquitoes. The results were really dramatic. All

[4]The original work was summarized by A.C. Allison in "Sickle cells and evolution," *Scientific American* 195(1956):87–94. Further studies were reported confirming the initial observations in "Recent developments in the study of the inherited anemias," *Eugenics Quarterly* 6(1959):155–66. Many investigators have confirmed and extended these pioneer studies.

fifteen inoculated natives with normal hemoglobin had severe attacks of malaria, which Allison then cured with antimalarial drugs. Of the fifteen persons with both sickle hemoglobin and normal hemoglobin (heterozygotes), only two displayed any demonstrable infection, and those two cases were very light. Later studies have adequately established the validity of this remarkable relationship between the kind of hemoglobin in the red cells and the resistance to malaria.

Throughout much of equatorial Africa, children who possess only normal hemoglobin frequently die of malaria, while children with only sickle hemoglobin die of the sickle cell anemia; and human life is chiefly carried on by survival to adulthood of the heterozygotes who possess both types of hemoglobin in roughly equal amounts. Other examples of this sort of relationship between some particular genetic constitution and some endemic disease of childhood importance are now known: hemoglobin C, thalassemia, etc.

It is also true that the heterozygote formed between two alleles present in a population may be selected *against.* This is true, for instance, of the now familiar hemolytic disease of the newborn caused by the Rhesus blood group incompatibility between fetus and mother. When a mother is Rh-negative and her offspring, because of the father's Rh-positive type, are Rh-positive, in about 5 percent of cases red blood cells from the fetus enter the maternal circulation and there stimulate the production of antibodies. These return into the fetal circulation, destroy its red blood cells, and thus cause the formerly severe or fatal disease. Whenever in a mixed population where both alleles are present there is fetal or infant loss on this account, the force of selection reduces the frequency of the rarer of the alleles and increases the frequency of the commoner one. We should consequently expect that in the course of time, in such a population as any of northern Europe or the United States, the rarer Rh-negative allele would disappear altogether and only the already commoner Rh-positive allele would persist, even as it does now in most Asian, American Indian, and Pacific peoples. But various complications may arise to prevent so simple and relatively quick an attainment of genetic uniformity.

A reduced penetrance is one of these mechanisms that retard the effect of selection. That is to say, a particular genotype may not always manifest itself or may do so only at a fairly advanced age. Thus the maternal-fetal Rh incompatibility never arises in about 95 percent of all Rh-negative mothers who bear Rh-positive children. For one thing, at least one Rh-positive fetus is required to initiate the maternal immunization, and that first case itself escapes harm. For another thing, most Rh-positive men are heterozygous for the Rh-negative allele, in which case, by chance, half their children by Rh-negative wives are expected to be Rh-negative and to escape any ill consequences; and in small families all the children may happen to be Rh-negative even when only half of them are expected to be of that type. But of the greatest importance, it would seem, is the fact that even when the mother is immunizable and the fetus is Rh-positive, the red cells of the fetus may be prevented from reaching the antibody-forming centers of the mother's body. This may be either because the placenta forms in many instances an impervious barrier, or because the red blood cells are rapidly destroyed upon entering the maternal circulation.

Huntington's chorea, a disorder of the nervous system that is seriously debilitating and commonly fatal, is well known to be inherited through a simple dominant gene. The age of onset of the disease, however, is generally after the age of 30 or 35 years. Hence most persons carrying this very detrimental gene appear quite normal during those years of life when they are most likely to marry and have a family. The penetrance of the condition is less than 100 percent because some carriers of the gene die of other causes before the usual age of onset, although after they may have transmitted it to some of their children.

Other factors may also interact with the purely biological ones. An example is the phenomenon of "compensation," which has been discovered to apply to a number of situations wherein hereditary factors are subjected to adverse selection. Some years ago I found that reproductive compensation might be operating among the white families of Baltimore subjected to the maternal-fetal Rh incompatibility. At any rate, although in that city Rh-negative Negro mothers clearly had fewer living children than Rh-positive mothers, it was not so among the white families of the population, in which the average family size was considerably smaller. Here the Rh-negative mothers, although experiencing more losses of children and more abortions late in pregnancy than Rh-positive mothers, nevertheless made up for that loss by additional pregnancies and succeeded in producing even more living children than the Rh-positive women. One might suppose that the loss of anticipated offspring had sharpened their desires to attain the average size of family accepted as a social pattern in their part of the population, and that this heightened motivation had actually led them somewhat to overshoot the mark. Similar relationships have been described for acholuric jaundice, a rather rare hereditary disorder, and for thalassemia, a hemoglobin disorder common in certain Mediterranean populations. It is, to be sure, not certain how general and how important reproductive compensation really is in determining the general outcome of competition between alleles in a population; but the point to be emphasized is that the situation is one of complex, interacting forces and is not readily deducible from the primary effects of the alleles.

Bernice Cohen has provided a fine example of the complexity of such interactions in studies of the interrelationships between the ABO and Rh blood group systems.[5] It is not so long since no one had any good evidence of the selective effects of the three main alleles of the ABO system. Some anthropologists thought of the ABO blood groups as fine examples of truly neutral genetic characters. Now it is evident that this is not so. The presence of antigen A, and to a lesser extent of antigen B, on red cells that may leak into the maternal circulation from the fetus will lead to the rapid destruction of those cells, or will at least prevent them from getting to the antibody-forming centers if the mother carries anti-A or anti-B antibodies in her blood—as she will especially if she is of type O. It follows that a maternal-fetal incompatibility with respect to A or B will protect the mother from sensitization by the Rh antigens, if there is simultaneously a maternal-fetal Rh incompatibility. Hence the ABO constitution of mother and offspring plays an important part in the action of selection on the Rh blood group

[5]Bernice H. Cohen and Bentley Glass, "The relation of the ABO and Rh blood groups to differential reproduction," *J. Med. Educ.* 34(1959):414–17. Also, "Further observations on the ABO blood groups and the sex ratio," *Amer. J. Human Genetics* 11(1959):274–78.

system. Moreover, the interaction has a "feed-back." Presence of a double incompatibility, for both ABO and Rh systems, is more favorable in outcome than presence of a single incompatibility, at least for the Rh system. The Rh constitution of the mother and fetus thus in certain portions of the population favors particular alleles at the ABO locus. It is consequently quite impossible to consider the overall pattern of selection with respect to either the Rh or the ABO blood groups without considering the other also.

GENE FLOW

Enough has been said up to this point to hint that the exchange of genes between populations—the process of gene flow, as the geneticists now call it—has had a very great role in the establishment of human racial differences. The great racial stocks must have arisen in a genetic isolation almost if not quite complete. That isolation was of course largely geographical, a consequence of man's primitive ways of travel. Even today there exist geographical and cultural isolates which play a part in the establishment and maintenance of these genetic differences. Studies of the frequency of different blood group genes have been made in some of the subcastes of Bombay and these studies show that the religious and social restrictions upon intermarriage imposed by the caste system unmistakably lead to significant genetic differences in the several isolates examined.[6] The blood group frequencies, in particular, reveal a distinctness in some instances as great as that distinguishing different European nationalities or, say, U.S. Whites and Negroes. A subsequent study of the morphological features of these groups has revealed that in these respects, too, the endogamous (inmarrying) subcastes are as genetically distinct as they are in respect to their blood groups. It is rather surprising that differences of such magnitude could arise following isolation—incomplete isolation, at that—for no more than a

[6]L.D. Sanghvi and V.R. Khanolkhar, "Data relating to seven genetical characters in six endogamous groups in Bombay," *Annals of Eugenics* 15(1949):52–64.

few thousand years. But there is other evidence to strengthen the conclusion that this may happen.

With my colleagues, Milton S. Sacks, Elsa Jahn, and Charles Hess, I some years ago surveyed the frequencies of several blood group systems and other inherited traits in a small community of Old German Baptist Brethren, or Dunkers, living in Franklin County, Pennsylvania.[7] Because of their strict religious observances, this religious community is almost entirely endogamous; that is to say, marriages to outsiders are limited to the relatively few converts to the sect. Very significant genetic differences were found to exist between the general white population of the eastern United States and the Dunkers, who are largely descended from emigrants in the early eighteenth century from the Rhineland. Thus, in the space of about two centuries the isolation imposed on the community by its religious observances has been great enough to prevent whatever gene flow there has been from the surrounding population into the community from making the Dunkers genetically similar to their neighbors.

L.C. Dunn found a similar degree of distinctive genetic character in the Roman Jewish community descended from the inhabitants of the ghetto maintained there until about a century ago.[8] This Roman Jewish community is remarkably distinctive in its possession of certain blood group genes and differs both from the Italian Catholic population and even from the non-Roman Jewish community now residing in Rome. A very high frequency of blood group B was found, and one of the elsewhere exceedingly rare Rh genes (r') attained a frequency of over 5 percent. Jews from Tripoli who have recently immigrated to Israel have been found to resemble these Roman Jews rather closely. In general, however, the Jews of different countries are not so much alike in these respects; instead they resemble the populations among which they have lived for centuries. Thus the "Black Jews" of Cochin, in India, resemble their Hindu

[7]Bentley Glass, Milton S. Sacks, Elsa F. Jahn, and Charles Hess, "Genetic drift in a religious isolate: An analysis of the causes of variation in blood group and other gene frequencies in a small population," *American Naturalist* 86(1952):145–60.

[8]L.C. Dunn, *Heredity and Evolution in Human Populations* (Cambridge, Mass.: Harvard University Press, 1959).

neighbors in their gene frequencies; the Jews from eastern and southern Europe and northern Africa are quite alike and resemble the Oriental Jews; the Yemenite Jews are like their Arab neighbors. These facts clearly show that the genetic isolation of religious endogamous communities from their neighbors is rarely complete, and that gene flow into the isolate from the general population tends in time to erase the differences between them.

Whenever gene flow persists for a long time in one predominant direction between populations that have initially possessed a considerable difference in the frequencies of certain alleles, then a geographical gradation in allelic frequencies results. This is called a cline. To be sure, clines can also result from a gradient of natural selection between one extreme of temperature or climate or some other physical agent and the opposite extreme; but in human evolution most of the clines noted and examined seem to rest upon other causes. The outward flow of the gene for tawny hair in the Australian aborigines from its center of origin and high frequency, as already described, results in clines radiating in all directions from that center. On a broader scale, there is a famous cline in the frequency of blood group B as one passes westward from its region of worldwide highest frequency, among the Buriat and Kalmuck Mongols, to the shores of the Atlantic (See Figure 5.4). Here the aboriginal condition is perhaps best shown by the Basque population of the northern coast of Spain and the southwestern coast of France. The Basques not only have a unique non-Indo-European language, they possess the lowest frequency of blood group B in Europe and the highest frequency of the Rh-negative allele (*r*) in the world. Across Turkestan, Russia, Poland, and Germany to France and the Basques there is a fairly regular decline in the frequency of the allele responsible for blood group B, except for pockets of high B frequency among the Magyars and Finns, who on other grounds are clearly the most Oriental of Europeans in ancestry.

Now we know of numerous historical invasions of Mongol hordes who swept into Europe from the east, pillaging, looting, burning, and raping. Perhaps there were even some of these before the time of Attila the Hun, who brought his hordes in

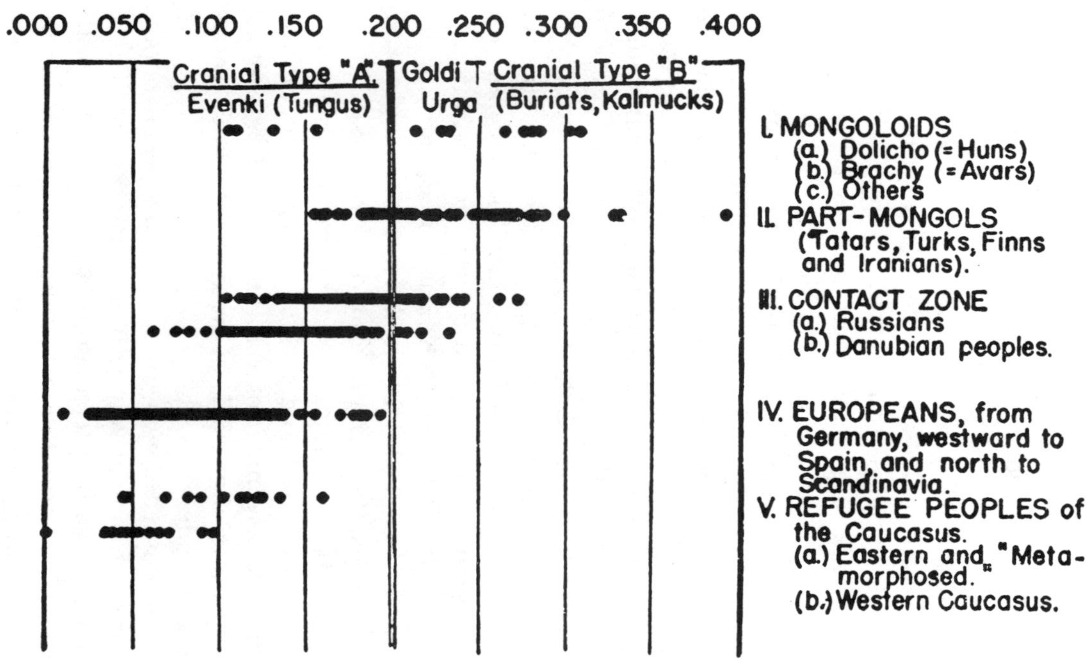

FIGURE 5.4. Frequencies of the blood group allele I^B in Asiatic and European populations. (From Boyd, after Candela).

the fifth century to the Rhine River, to the very ramparts of the Roman Empire. After him every century saw some similar incursion, although not all of them penetrated so far to the west. Until the time of Tamerlane, nearly a thousand years later, the wild inroads continued. Perhaps one should stop to consider whether all those Mongol invaders went peacefully home again after their conquests. At any rate, the evidence is entirely against it. They have left their genes behind them, as some Europeans and Americans bear witness, namely, all those who are of blood group B or AB. For there is many times as high a probability, for such persons, that the B gene (allele) was inherited from some forgotten Mongol ancestor as that it came from one of the rare European aborigines who had it. Let me hasten to add that, for every B gene transmitted to the present day from each Mongol invader of Europe there are some ten thousand other genes, less readily identified, also inherited from him; and in the course of fifteen to fifty generations those genes have undergone recombination after recombination until it is most improbable that any of us who are of European ancestry altogether lack Mongol genes in our composition. No, we have simply conveniently forgotten our mixed ancestry. We no longer tag particular genes as "Mongolian" in our genotypes, since all of us possess a few, and there is no possibility of erecting social distinctions on such a basis. The gene flow has completed its work; and while all the different sorts of genes are still here among us, the racial intermixture, especially in the United States, fast approaches a random character, so that in almost any hundred caucasoid persons chosen at random you will find the same percentage of A, B, AB, and O blood groups, of brown-eyed and blue-eyed, of tasters and nontasters of phenylthiocarbamide.

Here in the United States we have still, however, an opportunity to observe in a relatively early stage the process of gene flow between genetically different populations. As we all know, one segment of our population was brought here in slavery beginning over three centuries ago and continuing until 120 years past. The Negro citizens of the United States were derived largely from the tribes of west and equatorial Africa; and it is possible by comparing frequencies of the same allele in the present populations of those parts of Africa, in our North American Negroes, and in the North American Whites, to arrive at a fairly exact, scientific estimate of the extent to which gene flow from the White into the Negro population has modified the latter. In the 1950s, with my colleague C.C. Li, I undertook a study of this process; and more recently, with better data becoming available from Africa in respect to the blood groups there, it has been possible to refine our earlier estimates.[9] The R° allele of the Rh blood groups supplies the best basis for such an undertaking, because it has a very high frequency, one of about 60 percent, in the African tribes, while it is very uncommon, with a frequency of 2.8 percent, in the people of Europe and of the United States. The frequency of R° in the U.S. Whites is not significantly different from that in Europe, from which fact it is evident that no gene flow from the American Negro population into the White population has yet occurred of a magnitude sufficient to modify the latter. That is not very strange, since the White population has remained during these three centuries just about ten times the size of the Negro population, and it would take a far greater transfer of genes from the Negro population to the White to modify the latter detectably than in the reverse direction to observe a change in the Negro population. The frequency of the R° allele in the U.S. Negro population of New York and Baltimore—and these urban centers are fairly representative of all except a few real Negro isolates such as the Gullah Negroes of Charleston, South Carolina—is 43 to 44 percent. From these figures it may be calculated that the present amount of White admixture in the U.S. Negroes is 22.1 percent or 28.5 percent, depending on the technical assumptions made. One must emphasize that although these estimates are derived from the R° frequencies in the three populations, similar tests may be made utilizing other genes, provided their frequencies are sufficiently different in African Negroes and North American Whites. This has in fact been done for a number of other genes, and the results agree satisfactorily.

[9]Bentley Glass and C.C. Li, "The dynamics of racial intermixture—An analysis based on the American Negro," *Amer. J. Human Genetics* 5(1953):1–20.

Since approximately ten generations have elapsed since the admixture began, one can also, upon the assumption that the gene flow has been uniform over the generations, estimate that it has amounted to 2.5 percent or 3.3 percent per generation. This statement is equivalent to saying that in every generation this same percentage of genes in the socially defined Negro population has been freshly introduced into it from intermating with Whites—or that double that percentage of "Negro" babies born have had one White parent. No one can say at present, at least on genetic evidence, whether the rate of gene flow was greater before the Civil War than since, or the reverse; but if it were to continue unaltered at the same average rate, it would still take 75 to 98 generations, or 2,000 to 2,700 years, in round numbers, to reach a final equilibrium like that of the Mongol genes in Europe. Thus, if inevitable, the process of racial breakdown is very slow—although of course the process is quite likely to speed up as the final stages are approached and as gene flow in the opposite direction, from the Negro into the White population, becomes significant. I emphasize also that what the above estimates are based on is the gene flow resulting largely from illicit mating. The gene flow would be profoundly changed in amount if the social barriers to interracial mating began to break down rapidly.

The same kind of analysis can be applied to answer the question of whether or not the Negroes of the United States have any considerable admixture of American Indian genes, as Negroes in Brazil, for example, most certainly do.[10] The answer is unequivocal. None of the genes used for measuring the gene flow indicate that anything of the sort has occurred. There may indeed have been a few Indian ancestors of present-day "Negro" groups, but their genes have been too few in the total gene pool of the Negro population, as represented in Baltimore, to make any significant difference.

Gene flow studies of this sort have recently been conducted by Saldanha and others in Brazil, with highly interesting

[10]Bentley Glass, "On the unlikelihood of significant admixture of genes from the North American Indians in the present composition of the Negroes of the United States," *Amer. J. Human Genetics* 7(1955):368–85.

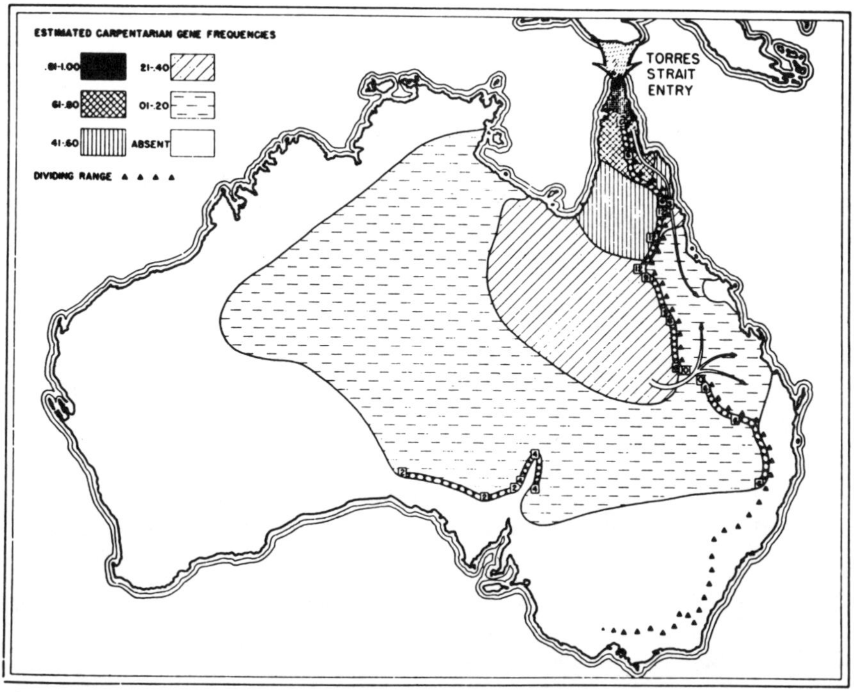

FIGURE 5.5. Gene flow model based upon the Torres Strait entry into Australia of the Carpentarian racial element. (From Birdsell).

results.[11] They have not yet, however, been applied to the population of our fiftieth state, which is so very interesting from a genetic standpoint because of this intermixture of Polynesian, Asiatic, and European racial strains of several sorts. It is not so easy to determine the proportions of the ingredients in a complex mixture as in a simple one of two components only; but Saldanha has shown that it may be done successfully. Data are being accumulated that will make the Hawaiian analysis possible.

Meanwhile the method has been applied with striking success to the analysis of the origin of the Australian aborigines.

[11] P.H. Saldanha, "Gene flow from White into Negro populations in Brazil," *Amer. J. Human Genetics* 9 (1957): 299–309.

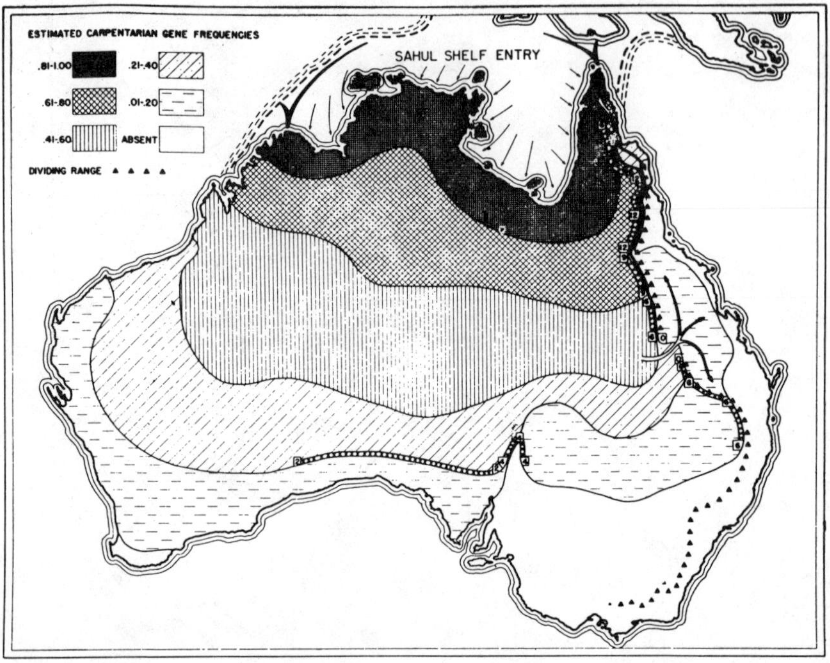

FIGURE 5.6. Gene flow model based upon entry of the Carpentarian element along the Sahul Shelf. (From Birdsell).

On various anthropological grounds it has been suspected that the earlier inhabitants of Australia were much later joined by a wave of rather different people, the Carpentarian stock. Where did these latecomers land? Two reasonable possibilities existed: either along the whole northern coast or across Torres Strait from New Guinea. Joseph Birdsell, an American anthropologist, has calculated what the distribution of gene frequencies would be in the two cases and then turned to an analysis of the actual distribution.[12] The answer was unequivocal. Entry by Torres Strait was ruled out, and entry along the northern Sahul Shelf vindicated, although on a more restricted length of coastline than had been supposed. (See Figures 5.5, 5.6, and 5.7.)

[12]Birdsell, op. cit., pp. 259–314.

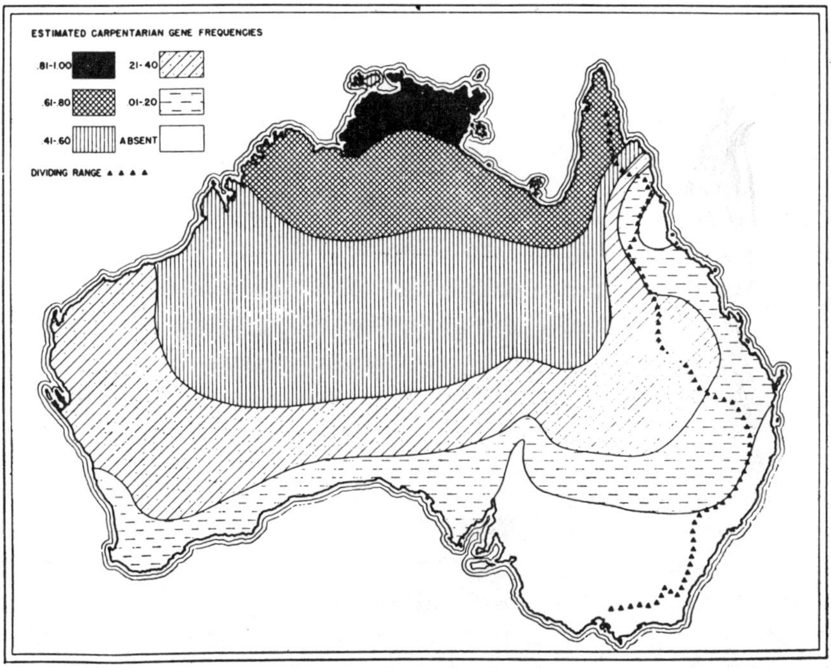

FIGURE 5.7. Actual distribution of the Carpentarian element in recent aboriginal Australian populations, as worked out by Birdsell. (From Birdsell).

RANDOM GENETIC DRIFT

The final factor that modifies gene frequencies is chance itself. This may operate in either of two ways. One we may call the "founder effect," the other "genetic drift" proper; but there is really no fundamental distinction between them, since in both cases the effect is the result of taking small samples of breeding groups that are very few in number of individuals. These are usually nonrepresentative of the populations from which they are drawn. The first, or founder, effect arises when a few individuals from a large population are drawn off to start a new population. In flies, these might be the few survivors in springtime of the enormous populations of the preceding summer. In our own species, they might be the few emigrants who populate a new land. In either case, the few individuals who propagate a new population cannot fully and in the right proportions represent the numerous genes and all the allelic frequencies of the ancestral population. Thus, in our Dunker studies, we found that certain distinctions of the Dunker community, especially the almost complete absence of the allele for blood group B and the high frequency of the allele for blood group A, dated back at least three generations. It is readily conceivable that the few original immigrants who started the Dunker community in Franklin County, Pennsylvania merely by chance failed to include any persons of group B.

Secondly, in a population of a given genetic composition, if it is very small, each new generation will tend to deviate somewhat from the exact frequencies present in the parent generation; and the smaller the population, the greater these deviations are likely to be. Thus, in a population of 100 persons per generation comprising 200 genes at the MN blood group locus, 55 percent might be M and 45 percent N; but in the following generation there might be 65 percent M and 35 percent N, simply because of a random deviation from expectation in the fertilization of so few eggs by so few sperms. And in a third generation, with equal probability, one might get a return to the original 55 percent M, 45 percent N or a further reduction in the frequency of N: say to 73 percent M, 27 percent N. If the latter were to occur, it would represent *drift*. Like a run of luck, this might continue until one allele was completely eliminated from the population. The example is an actual one, the figures chosen being exactly what was found in the Dunker community.[13]

But one might say: "Well, what is the importance of random genetic drift in evolution? Populations are large and chances even out in the long run." Yet we must remember the past of our species as well as its present. Before the dawn of agriculture, a mere seven to ten thousand years ago, *Homo sapiens* lived for immemorial generations in small tribes, of perhaps a few hundred persons, and often, I would suppose, of even fewer. Under just such conditions chance would play a great part in

[13]Bentley Glass, "On the evidence of random genetic drift in human populations," *Amer. J. Physical Anthropology*, n.s., 14(1956):541–55.

determining what genes, provided they were not *too* harmful, would persist in particular tribes. (Also, because of the small numbers of individuals, a mutation that today can be found many times over in a city the size of Chicago [$10^{-5} \times 4 \times 10^{6} = 40$] would not occur more than once in 500 generations, or 12,500 years, in a tribe of 100 persons [$10^{-5} \times 2 \times 10^{2} = 2 \times 10^{-3}$]. So whatever mutation occurred in one tribe would be highly unlikely to occur in another.) Birdsell and his colleagues in Australia have found not only an apparently single occurrence of the mutation to tawny hair, but also abrupt changes in the frequencies of the blood group genes from isolate to isolate, while at the same time they show a cline from north to south. The pattern is most simply explained as involving natural selection, to explain the cline, and random genetic drift, to explain the abrupt and unpredictable differences between adjacent isolates. Arthur Steinberg has found erratic differences in the gene frequencies of small Hutterite communities (another religious isolate in the United States). Gabriel Lasker has calculated that genetic drift may amount to 1 to 8 percent in the rural towns of modern Mexico, in which case it ought often to be sufficient to constitute the determining element in genetic change, rather than selection.[14] In short, random genetic drift may well have played a very great part in the development of the *initial* tribal differences of mankind. And then, as some few of these tribes acquired agriculture and mushroomed in size, the initial genetic composition may well have set the character of an entire geographic race.

How may we sum up these interesting but intricate involvements of evolutionary factors in the molding of mankind? It seems, in the first place, that genetic differences seen in the human species are of a minor, not a major, character. We are indubitably of one human species, and more and more rapidly today the racial distinctions are melting away, as those of past time have sometimes done. But it is most unlikely that renewed isolation will arise and provide the requisite conditions for the generation of any new races of mankind—unless a nuclear war were to scatter the remnants of humanity in remote regions and

[14]G.W. Lasker, "Human evolution in contemporary communities," *Southwestern J. Anthropology* 10(1954):353–65.

unless civilization was reduced to the level of barbarism. The genetic differences which do distinguish races are relatively few at the morphological level of skin color, eye color, hair type, stature, conformation of head and body, and the like. They are more numerous at the physiological and biochemical level, but here, as at the morphological level, racial genetic differences are rarely all-or-none; usually they are to be described only in terms of more common or less common alleles. In this respect gene flow tends to establish an equilibrium representing a homogeneous distribution. Some day we may see, throughout the world, innumerable different genetic sorts of individuals but no strikingly different genetic sorts of populations. Clearly, however, that will not come to pass for a long time hence, perhaps several thousands of years. And although such a timespan is still a short one in the history of our species, the incredibly rapid advance of technology in the age of science will not wait upon slow solutions.

Some of our racial genetic differences may have arisen by chance. If so, they should have for us no real importance. Others may be the product of selection acting in different environments. If so, each selected gene is superior in its own setting. We have also learned that even the harmful gene may be necessary to the preservation of life in certain adverse environments, though when we correct the environment it may become possible to rid the population of the no longer necessary gene.

Perhaps, in the end, what these things should teach us is mutual respect. There may still be prejudice, hatred, and fear when there is no more race. Let us hope that we have sufficient time to learn from race, before it vanishes from humankind, the lesson of mutual respect.

6 The Ethical Basis of Science

It has been said that science has no ethical basis, that it is no more than a cold, impersonal way of arriving at the objective truth about natural phenomena. This view I wish to challenge, since it is my belief that by examining critically the nature, origins, and methods of science we may logically arrive at a conclusion that science is ineluctably involved in questions of values, is inescapably committed to standards of right and wrong, and unavoidably moves in the large toward social aims.

Human values have themselves evolved. Man arose after some two billion years of organic evolution, during which species after species originated, flourished, and fell or occasionally became the progenitors of species that were new and better adapted, on the basis of the evolutionary scheme of values. Fitness, like it or not, in the long run meant simply the contribution of each trait and its underlying genes to survival. High mortality or sterility led to extinction; good viability and fertility enabled a gene or a trait, an individual or a species, to be perpetuated. Man's own values grew out of his evolutionary origins and his struggle against a hostile environment for survival. The loss of certain unnecessary structures, such as bodily hair once clothing was invented; the homeostatic regulation of body temperature and blood pressure, breathing, and

---

This article is a condensation from the book *Science and Ethical Values* (Chapel Hill: University of North Carolina Press, 1965). It was reprinted in *Science* 150 (1965): 1254–61 (1965) and is now reprinted by permission.

predominant direction of blood flow; embryonic and fetal growth inside the mother and the prolonged dependence upon maternal lactation; the slow maturation that enabled the brain to enlarge so greatly; the keen vision so necessary to the hunter using his weapons, and many other important human characteristics that contributed to the social nature of man and cemented the bonds of family and tribe arose adventitiously, were improved step by step, and endured because they promoted human survival. Our highest ethical values—the love of the mother for her child and of the man for his mate, the willingness to sacrifice one's own life for the safety of the family or tribe, and the impulse to care for the weak, the suffering, the helpless—had the same primitive beginnings.

But these ethical values are always, in the evolutionary scheme of things, relative and never absolute. Whenever the environment becomes changed, the adaptiveness of existing traits becomes maladjusted, and the forces of natural selection lead to a realignment of the genotype, an alteration of the external features and modes of behavior, a modification of the species. What was once good is so no longer. Something else, in terms of reproductive fitness, has become better.

Finally, a crude, embryonic form of science entered the scheme of things, a method of observing and reporting accurately to other persons the movements of the stars, the planets, and the sun and moon, the behavior and migrations of the food animals, the usefulness of certain seeds for food and of certain stems for fibers, the poisonous properties of others. For generations all such practical lore was transmitted only by word of mouth, but the day came when the useful knowledge could be written down and preserved inviolate from the forgetfulness and the twists of memory. These were the first simple steps in the development of science: observation, reporting, written records, communication. To such must be added the processes of human reasoning, at first mostly by analogy, so often wrong, then by improved analysis, by deduction from an established truth, or by induction of an established truth from a multitude of observations.

Seen aright, science is more than the instrument of man's increasing power and progress. It is also an instrument, the finest yet developed in the evolution of any species, for the malleable adaptation of man to his environment and the adjustment of his environment to man. If the human species is to remain successful, this instrument must be used more and more to control the nature and the rate of social and technological change, as well as to promote it. In this sense, at least, science is far more than a new sense organ for comprehending the real relations of natural phenomena and the regularities we call "laws of nature." It is also man's means of adjustment to nature, man's instrument for the creation of an ideal environment. Since it is preeminently an achievement of social man, its primary function is not simply that of appeasing the individual scientist's curiosity about his environment—on the contrary it is that of adjusting man to man and of adjusting social groups in their entirety to nature, to both the constraints and the resources of the human environment.

Ethics is a philosophy of morals, a moral system that defines duty and labels conduct as right or wrong, better or worse. The evolutionist is quite prepared to admit the existence of right and wrong in terms of the simple functions of biological structures and processes. The eye is for seeing, an evolutionary adaptation that enables an animal to perceive objects at a distance by means of reflected light rays. Sight conveys information about food, water, danger, companionship, mating, the whereabouts and doings of the young ones, and other vitally important matters. Should one not then say, "To see is right; not to see is wrong"? Similarly, the mind reasons as it does because in the countless ages of evolutionary development its characteristic mental processes led to successful coping with the exigencies of life. Humans whose mental processes, because of different genes, too often led them to wildly erroneous conclusions did not so often leave children to reason in similar ways. It is thus right to be guided by reason, wrong to distrust it. Does it not follow, finally, from consideration of the social role and function of science, that it is *right* to utilize science to help develop and regulate human social life, adjustment to change, and rate of social transformation? Conversely, it is *wrong*—morally and ethically wrong—not to do so. This does not imply that science should be, or can be, our sole guide. We must use whatever light and whatever reason we have to chart our course into the unknown.

Those who distrust science as a guide to conduct, whether individual or social, seem to overlook its pragmatic nature, or perhaps they scorn it for that very reason. Rightly understood, science can point out to us only probabilities of varying degrees of certainty. So, of course, do our eyes and ears, and so does our reason. What science can do for us that otherwise we may be too blind or selfwilled to recognize is to help us to see that what is right enough for the individual may be wrong for him as a member of a social group, such as a family, that what is right for the family may be wrong for the nation, and that what is right for the nation may be wrong for the great brotherhood of man. Nor should one stop at that point. Man as a species is a member—only one of the members—of a terrestrial community and an even greater totality of life upon earth. Ultimately, what is right for man is what is right for the entire community of life on earth. If he wrecks that community, he destroys his own livelihood. In this sense, coexistence is not only necessary but also right, and science can reveal to us the best ways to harbor our resources and to exploit our opportunities wisely.

THE SUBJECTIVITY OF SCIENCE

From the foregoing description of science as itself an evolutionary product and a human organ produced by natural selection, it may already be guessed that I do not adhere to the view that either the processes or the concepts of science are strictly objective. They are as objective as man knows how to make them, that is true; but man is a creature of evolution, and science is only his way of looking at nature. As long as science is a *human* activity, carried on by individual persons or by groups, it must at bottom remain inescapably subjective.

Our sensory apparatus and the structure of the human nervous system, within which arise our sensations, grow and develop as they do from the first beginnings in the human embryo because of the particular genetic constitutions we inherit from our parents. First and foremost, we are *human* scientists, not insect scientists, nor even monkey scientists. The long past of our evolutionary history, with its countless selections and rejections of various kinds of genes and combinations of genes, has made us what we are. Try as we will, we cannot break the bonds of our subjective interpretations of the physical events of nature. We are born blind to many realities and at best can apprehend them only by translating them by means of our instruments into something we can sense with our eyes or ears, into something we can then begin to reason about by developing abstract mental concepts about them, by making predictions on the basis of our hypotheses, and by testing our theories to see whether reality conforms to our notions.

This line of reasoning leads us to the conclusion that the objectivity of science depends wholly upon the ability of different observers to agree about their data and their processes of thought. About quantitative measurements and deductive reasoning there is usually little dispute. Qualitative experiences like color, or inductive and theoretical types of reasoning, leave great room for disagreement. Usually, they can be reduced to scientific treatment only if the subjective color can by agreement be translated into some quantitative measurement such as wavelength, only if the reasoning can be rendered quantitative by use of a calculus of probability. It nevertheless remains a basic fact of human existence that the subjectivity of the individual personality cannot be escaped. We differ in our genes, each of us possessing a genotype unique throughout all past and future human history (unless we happen to possess an identical twin). To the extent that our genes endow us with similar, though not identical, sensory capacities and nervous systems, we may make similar scientific observations, and we may agree to ignore the existence of the variables in our natures that prevent us from ever making exactly the same measurements as someone else or arriving at exactly the same conclusions. But it is perilous to forget our genetic individuality and our own uniqueness of experience. These form the basis of the ineradicable subjectivity of science. In the last analysis, science is the common fund of agreement between individual interpretations of nature. What science has done is to refine and extend the methods of attaining agreement. It has not banished the place of the individual observer, experimenter, or theoretician, whose work is perhaps subjective quite as much as objective.

These considerations may seem so obvious as not to require the emphasis just given them. Yet I believe not. Somehow there has crept into our writings about the nature and methods of science a dictum that science is objective while the humanistic studies are subjective, that science stands outside the nature of man. What a profound mistake! Science is ultimately as subjective as all other human knowledge, since it resides in the mind and the senses of the unique individual person. It is constrained by the present evolutionary state of man, by the limitations of his senses and the even more significant limitations of his powers of reason. All that can be claimed for science is that it focuses upon those primary observations about which human observers (most of them) can agree, and that it emphasizes those methods of reasoning which, from empirical results or the successful fulfillment of predictions, most often lead to mental constructs and conceptual schemes that satisfy all the requirements of the known phenomena.

SCIENCE, INTEGRITY, AND INTELLECTUAL FREEDOM

From a consideration that science is a human activity, inescapably subjective, and a product of biological evolution, it is possible to derive a genuine ethical basis of science. J. Bronowski, in an essay entitled "The Sense of Human Dignity,"[1] has sketched a treatment that serves well for a beginning. The values and duties which are the concern of ethics are social, he affirms. The duties of men hold a society together, he says, and "the problem of values arises only when men try to fit together their need to be social animals with their need to be free men." Philosophy must deal with both the social and individual aspects of value. Most philosophical systems have found this very difficult to do. Thus dialectical materialism swings far to the side of social values and leaves little scope for individual freedom. Positivism and analytic philosophy, as typified by Bertrand Russell and Wittgenstein, on the other hand, emphasize the values of the individual.

Hence, continues Bronowski, because the unit of the positivist or the analyst is one person, "positivists and analysts alike

[1] J. Bronowski, *Science and Human Values* (New York: Messner 1956), p. 63.

believe that the words *is* and *ought* belong to different worlds, so that sentences constructed with *is* usually have a verifiable meaning, but sentences constructed with *ought* never have."[2]

The issue, then, is simply whether verification can indeed be assumed to be carried out by one man. Bronowski concludes, and I find it impossible to deny, that in the practice of science this supposition is sheer nonsense. Verification depends completely on the existence of records that may be consulted, of instruments that may be used, of concepts that must be understood and be properly utilized. In all these ways, knowledge is a social construct, science a collective human enterprise, and verification is no procedure of the naked, unlettered, resourceless man but an application of the collective tools of the trade and the practiced logic of science to the matter at hand. It is a fallacy to assume that one can test what is true and what is false unaided. But then it must follow that all verification, all science, depends upon communication with others and reliance upon others. Thus we come straight to the *ought* of science, for we must be able to trust the word of others. A full and true report is the hallmark of the scientist, a report as accurate and faithful as he can make it in every detail. The process of verification depends upon the ability of another scientist, of any other scientist who wishes to, to repeat a procedure and to confirm an observation.

Neither the philosophy of dialectical materialism nor that of the individualist accords with the basic nature of man and of scientific truth. The extreme social position leaves no room for the conscience of man and the exercise of intellectual freedom because the community dictates what is right and what a man *ought* to do. Yet the positivist's position is also faulty because "how a man *ought* to behave is a social question, which always involves several people; and if he accepts no evidence and no judgment except his own, he has no tools with which to frame an answer."[3] Again, "All this knowledge, all our knowledge, has been built up communally; there would be no astrophysics, there would be no history, there would not even be language, if man were a solitary animal."[4] "What follows?" asks Bronowski,

[2]Ibid., p. 72
[3]Ibid.
[4]Ibid., p. 73

and he answers: "It follows that we must be able to rely on other people; we must be able to trust their word. That is, it follows that there is a principle which binds society together, because without it the individual would be helpless to tell the true from the false. This principle is truthfulness. If we accept truth as a criterion, then we have also to make it the cement to hold society together." Whence he derives the social axiom:

> We OUGHT to act in such a way that what IS true can be verified to be so.[5]

So Bronowski. If his reasoning be accepted, and to me it seems unarguable, we must conclude that the cement of society is nothing less than the basic ethical tenet of science itself. The very possibility of verification, the assurance that one's own conclusions are not dreams, hallucinations, or delusions rests upon confirmation by others, by "competent" observers whom we trust to tell the truth.

The Scientist's Integrity

Ethics rests upon moral integrity. Science rests upon the scientist's integrity. This is so implicit in all of our science that it is rarely expressed and may be overlooked by the novice or layman. Bronowski mentions examples of what happens when this basic moral commandment is violated by a scientist. Lysenko is held up to scorn throughout the world and eventually is deposed.[6] Kammerer commits suicide.[7] It is very interesting that both of these notorious examples, and others less well known, such as that of Tower, a quondam professor of biology at the University of Chicago, have related to attempts to "prove" or bolster the theory of the inheritance of acquired characteristics. The singular attractiveness of this theory for violators of scientific integrity is no doubt owing to its social significance, since if true it would offer a quick and easy way for man to

[5]Ibid.

[6]B. Glass, "Dialectical materialism and scientific research," *Quart. Rev. Biol.* 23 (1948): 333–37; D.I. Greenberg, "Lysenko: Soviet science writes finis to geneticist's domination of nation's biological research," *Science* 147 (1965): 716–17.

[7]R.B. Goldschmidt, "Research and politics," *Science* 109 (1949): 219–27.

control the direction of human evolution and would lessen the obdurate qualities of genes modifiable only by mutation in uncontrollable directions.

It is not so generally recognized by these superficial evolutionary philosophers that, if true, the inheritance of characters produced through modifications of the environment would call in question the value of all evolutionary gains, since the modified characters would themselves have no real genetic permanence and would shift and vary with every change of environment. They also do not recognize one of the most essential aspects of heredity, the protection of the genetic nature against vicissitudes. The reason why death is so necessary a part of life is that the ground must be cleared for fresh life. The reason why the genotype must remain unmodifiable by ordinary environmental causes is because the course of life for every individual involves the cumulative effects of injury, disease, and senescence. The new generation must indeed start *fresh*—that is, free from all the disabilities incurred during life by its parents and earlier ancestors. Evolution through the action of natural selection upon mutations, most of which are harmful and nonadaptive, while only a rare exemplar among them is possibly advantageous, is a process slow in the extreme. But it preserves the gains of the past, and it permits every generation to be born anew, unburdened by decrepitude, to try out its varieties of genotypes in each niche of the environment.

The loss of scientific integrity through deliberate charlatanry or deception is less common than the violation of scholarly honesty through plagiarism. The theft of another man's ideas and the claim that another's discovery is one's own may do no injury to the body of scientific knowledge, if the substance of what is stolen be true. It may even do no harm to the original discoverer, who may be dead or in no need of further credit to advance his own career. It is nevertheless a canker in the spirit of the thief and does damage to the fabric of science by rendering less trustworthy the witness of the scientist.

Plagiarism shades into unacknowledged borrowing. Which of us in fact can render exactly the sources of all his ideas? Psychologists have now amply demonstrated the ease with which self-deception enters into the forgetfulness of borrowed benefits. The wintry wind of man's ingratitude blows only on the donor of benefits forgot. Around the self-deluded recipient blow only the mildest, gentlest zephyrs of spring. The newer patterns of scientific publication and support of research have multiplied a thousandfold the opportunities for the scientist's self-deception. Editors of scientific journals today customarily rely upon referees for opinions regarding the merit of manuscripts submitted for publication. The enormous expansion of scientific activity and the development of hundreds of new specialties have made this referee system necessary. The best referee is of course some other scientist who is working closely on the same scientific problems but is not associated with the author in the actual work—in other words, a competitor, since we must not forget that scientists are people who must earn a living, and since compensation and repute follow productivity and publication. Natural selection is at work among scientists, too! What is most alarming about the workings of the referee system is not the occasional overt lapse of honesty on the part of some referee who suppresses prompt publication of a rival's work while he harvests the fruit by quickly repeating the work—or even extending it—and rushing into publication with his own account. What is far more dangerous, I believe, because it is far more insidious and widespread, is the inevitable subconscious germination in the mind of any referee of the ideas he has obtained from the unpublished work of another person. If we are frank with ourselves, none of us can really state where most of the seminal ideas that lead us to a particular theory or line of investigation have been derived. Darwin frankly acknowledged the ideas of Malthus which led him to the theory of natural selection: but although he was one of the most honest of men, and one who was deeply troubled when Alfred Russel Wallace sent him in 1858 the brief paper setting forth his own parallel derivation of Darwin's theory, Darwin nevertheless never made the slightest acknowledgment of the idea of natural selection which he had surely read in the work of Edward Blyth in 1835 and 1837.[8] We may guess that Darwin's reasoning at the time went rather as follows:

> Blyth's conception is that natural selection leads to a restriction of hereditary variation in populations. Through

[8]L.C. Eiseley, "Charles Darwin, Edward Blyth, and the Theory of Natural Selection," *Proc. Amer. Phil. Soc.* 103 (1959): 94–158.

> elimination of the more variable specimens in a species, nature keeps the species true to type and prevents it from becoming maladapted to its environment. Blyth's "natural selection" is not an evolutionary force at all, but instead a force for the maintenance of the status quo.

Yet it is very hard to understand why, when the full significance of the action of natural selection dawned upon Darwin, he did not re-examine the ideas of Edward Blyth. It should have been perfectly evident to him that the very same force that would eliminate variation and maintain the status quo of the species in a stationary environment would operate quite differently in a changing environment. Will we then ever know the extent to which Darwin was really indebted to Blyth, or how the ideas he probably rejected as invalid actually prepared the way for his reception of Malthus's thoughts in 1838?

The conscientious referee of unpublished scientific manuscripts is similarly a gleaner in the harvest fields of others. The only possible way to avoid taking an unfair advantage would be to refuse to referee any manuscripts that might conceivably have a relationship to one's own research work. The consequences for editors left with piles of unevaluated manuscripts might become desperate, were there not, as I believe, a reasonable solution in the possibility that the role of referee could be limited to scientists who have ceased to do active experimental work themselves. What with the increasing life span and the large number of retired but mentally vigorous older scientists, the supply of competent referees would perhaps be sufficient. To be sure, the criticism may be raised that the older scientific men cannot properly evaluate the significance and merit of really revolutionary new ideas and lines of work. Neither, for the most part, can the young! A combination of older referees in the field and younger ones knowledgeable but not working in the same specialty might solve this difficulty.

What has been said about referees applies with even greater force to the scientists who sit on panels that judge the merit of research proposals made to government agencies or to foundations. The amount of confidential information directly applicable to a person's own line of work acquired in this way in the course of several years staggers the imagination. The most conscientious man in the world cannot forget all this, although he too easily forgets when and where a particular idea came to him. This information consists not only of reports of what has been done in the recent past but of what is still unpublished. It includes also the plans and protocols of work still to be performed, the truly germinal ideas that may occupy a scientist for years to come. After serving for some years on such panels I have reached the conclusion that this form of exposure is most unwise. One simply cannot any longer distinguish between what one properly knows, on the basis of published scientific information, and what one has gleaned from privileged documents. The end of this road is self-deception on the one hand or conscious deception on the other, since in time scientists who must make research proposals learn that it is better not to reveal what they really intend to do, or to set down in plain language their choicest formulations of experimental planning. Instead, they write up as the program of their future work what they have in fact already performed. Again, the integrity of science is seriously compromised.

Science and Intellectual Freedom

The first commandment in the ethical basis of science is complete truthfulness, and the second is like unto it:

> Thou shalt not covet thy neighbor's ideas nor steal his experiments.

The third is somewhat different. It requires fearlessness in the defense of intellectual freedom, for science cannot prosper where there is constraint upon daring thinking, where society dictates what experiments may be conducted, or where the statement of one's conclusions may lead to loss of livelihood, imprisonment, or even death.

This is a hard ethic to live by. It brought Giordano Bruno to the stake in 1600. The recantation of Galileo was an easier way; the timidity of Descartes and Leibniz, who left unpublished their more daring scientific thoughts, was understandably human but even less in the interest of science or, ultimately, of the society that felt itself threatened. Whether in the conflict of science with religion or with political doctrine (as in Nazi

Germany), or with social dogma (as in the Marxist countries), scientists must be willing to withstand attack and vilification, ostracism and punishment, or science will wither away and society itself, in the end, will be the loser.

From the beginning the inveterate foe of scientific inquiry has been authority—the authority of tradition, of religion, or of the state—since science can accept no dogma within the sphere of its investigations. No doors must be barred to its inquiries, except by reason of its own limitations. It is the essence of the scientific mind not only to be curious but likewise to be skeptical and critical—to suspend judgment until the facts are in, to be willing always, in the light of fresh knowledge, to change one's conclusions. Not even the "laws" of science are irrevocable decrees. They are mere summaries of observed phenomena, ever subject to revision. These laws and concepts remain testable and challengeable. Science is thus wholly dependent upon freedom—freedom of inquiry and freedom of opinion.

But what is the value of science to man, that it should merit freedom? There are those, indeed, who say that science has value only in serving our material wants. To quote one of them: "Science is a social phenomenon, and like every other social phenomenon is limited by the injury or benefit it confers on the community.... The idea of free and unfettered science ... is absurd." Those were the words of Adolf Hitler, as reported by Hermann Rauschning.[9] In Soviet states a similar view is held officially; and in the Western democracies, likewise, not a few scientists as well as laymen have upheld a similar opinion. The British biologist John R. Baker has pointed out that this view shades through others, such as the admission that scientists work best if they enjoy their work, and the supposition that science has value in broadening the outlook and purging the mind of pettiness, to the view that a positive and primary value of science lies in its creative aspect "as an end in itself, like music, art, and literature."[10] "Science aims at knowledge, not utility," says Albert Szent-Györgi;[11] and Alexander von Hum-

[9]H. Rauschning, *Hitler Speaks: A Series of Political Conversations with Adolf Hitler on His Real Aims* (London, Butterworths: 1939), pp. 220–21.

[10]J.R.Baker, *Science and the Planned State* (New York: Macmillan, 1945).

[11]A. Szent-Györgi, *World Digest* 55 (1943):50.

boldt wrote in his masterpiece, *Cosmos,* that "other interests, besides the material wants of life, occupy the minds of men."[12]

It is readily demonstrated that the social usefulness of the conclusions of science can rarely be predicted when the work is planned or even after the basic discoveries have been made. John R. Baker, in his book *Science and the Planned State,*[13] has cited numerous examples that show the impracticability of too narrowly planned a program of scientific research. The sphere of investigation must be determined by the investigator's choice rather than by compulsion—by perception of a problem to be solved rather than by a dogma to be accepted blindly. Science must be free to question and investigate any matter within the scope of its methods and to hold and state whatever conclusions are reached on the basis of the evidence—or it will perish. But science is represented only by the individual scientists. These persons must acknowledge the moral imperative to defend the freedom of science at any cost to themselves. Every Darwin needs a Thomas Henry Huxley. Every Lysenko demands his martyred Vavilov, his hundreds of displaced geneticists before he is finally deposed. Modern science, from its very beginnings near the end of the sixteenth century, became immediately concerned with a major political issue, the freedom of the scientist to pursue the truth wherever it might lead him, even though that conclusion might be highly disturbing to settled religious beliefs or social conventions and practice. The pyre of Bruno and the ordeal of Galileo led directly in spirit to the attacks on Charles Darwin 250 years later and to latter-day instances of the social suppression of scientific findings. The distortion of genetics by racists in Nazi Germany finds a counterpart in the United States. Mendelian genetics in the USSR and the nutritive qualities of oleomargarine in Wisconsin shared a similar fate—suppression by government. The third commandment then reads:

> Thou shalt defend the freedom of scientific investigation and the freedom of publication of scientific opinion with thy life, if need be.

[12]A. von Humboldt, *Cosmos: A Sketch of a Physical Description of the Universe,* E.C. Otte, trans. (London: Bohn, 1849).

[13]Baker, op. cit.

Science and Communication

Inasmuch as science is intrinsically a social activity and not a solitary pleasure, another primary aspect of the ethics of science is the communication to the world at large, and to other scientists in particular, of what one observes and what one concludes. Both the international scope of scientific activity and the cumulative nature of scientific knowledge lay upon the individual scientist an overwhelming debt to his colleagues and his forerunners. The least he can do in return, unless he is an ingrate, is freely to make his own contributions a part of the swelling flood of scientific information available to all the world.

There are at least five distinct obligations his indebtedness places upon each scientist. The first of these is the obligation to publish his methods and his results so clearly and in such detail that another may confirm and extend his work. The pettiness and jealousy that lead some scientists, in their effort to stay ahead of the ruck, to withhold some significant step of procedure or some result essential to full understanding of the stated conclusions have no place in the realm of science. In other instances it is sheer laziness or procrastination that is at fault. Whatever the only-too-human reason, science suffers.

A second obligation that is far more frequently neglected is the obligation to see that one's contributions are properly abstracted and indexed, and thus made readily available to workers everywhere. Many scientists ignore this obligation completely. Yet, as the sheer volume of scientific publication passes a half-million and soon a million articles a year, it is obviously insufficient to add one's own leaflet to the mountains of paper cramming the scientific libraries of the world. The need to have scientific findings abstracted and indexed has been fully recognized by such international bodies as the International Council of Scientific Unions: its Abstracting Board has urged every author to prepare an abstract in concise, informative style, to be printed at the head of each scientific paper; and the editors of most scientific journals have now made this a requirement for acceptance and publication of a paper. Nevertheless, few authors prepare their abstracts without a reminder, and few heed the requirements for a concise, informative summary that will permit proper indexing of the major items treated in the paper.

A third obligation is that of writing critical reviews, which will be true syntheses of the knowledge accumulating in some field. I firmly believe that there is no scientific activity today more necessary and at the same time less frequently well done than this one. I have said elsewhere:

> To be sure, the scientist seeks for facts—or better, he starts with observations.... But I would say that the real scientist, if not the scholar in general, is no quarryman, but is precisely and exactly a builder—a builder of facts and observations into conceptual schemes and intellectual models that attempt to present the realities of nature. It is the defect and very imperfection of the scientist that so often he fails to build a coherent and beautiful structure of his work....
>
> The creativity of scientific writing lies precisely here. The task of the writer of a critical review and synthesis... is not only indispensable to scientific advance—it surely constitutes the essence of the scientific endeavor to be no mere quarryman but in some measure a creator of truth and understanding. The aesthetic element that makes scientist akin to poet and artist is expressed primarily in this broader activity.
>
> The critical nature of the critical review grows from our constant forgetfulness of all this. The young scientist is taught carefully and methodically to be a quarryman or a bricklayer. He learns to use his tools well but not to enlarge his perspective, develop his critical powers, or enhance his skill in communication. The older scientist is too often overwhelmed by detail, or forced by the competition of the professional game to stick to the processes of "original research" and "training." The vastness of the scientific literature makes the search for general comprehension and perception of new relationships and possibilities every day more arduous. The editor of the critical review journal finds every year a growing reluctance on the part of the best qualified scientists to devote the necessary time and energy to this task. Often it falls by default to the journeyman of modest talent, a compiler rather than critic and creator, who enriches the scientific literature with a fresh molehill in which later compilers may burrow.[14]

[14]B. Glass, "The critical state of the critical review article," *Quart. Rev. Biol.* 39 (1964): 182–85.

All this need not be so, but it will remain so without a deeper sense of the obligation of the scientist to synthesize and present his broadest understanding of his own field of knowledge. Tomorrow's science stands on the shoulders of those who have done so, no less than on the shoulders of the great discoverers.

A fourth obligation is communication to the general public of the great new revelations of science, the important advances, the noble synthesis of scientific knowledge. There have always been a few eminent scientists who did not scorn to do this: Thomas Henry Huxley, John Tyndall, and Louis Pasteur set the pattern in the nineteenth century, and in our own time there have indeed been many who followed their precedent. Yet there seems to be a growing tendency to turn this obligation over to professional science writers who, however good, should not replace the direct, personal, and authoritative appeal of the scientist to the general public. As our culture and civilization become day by day more completely based on scientific discovery and technological application, as human exploration becomes ever more restricted to the endless frontiers of science, every citizen must know whereby he lives and whereupon he leans. A democracy rests secure only upon a basis of enlightened citizens who have imbibed the spirit of science and who comprehend its nature as well as its fruits. In fulfilling the requirement of our age for the public understanding of science, the scientist must shirk no duty.

A final obligation in the total purview of scientific communication is the obligation to transmit the best and fullest of our scientific knowledge to each succeeding generation. It is well said that genetic transmission of human characteristics and power is now far overshadowed by cultural inheritance. The transmission of knowledge is the role of the teacher, and the obligation of the scientist to teach is his last and highest obligation to the society that gives him opportunity to achieve his goals.

> To every scientist—to some sooner, to some only late—there comes the realization that one lifetime is too short and that other hands and other minds must carry on and complete the work. Only a few scientists are therefore content to limit their entire energies to exploration and discovery. Research

is one end, but the other must be the training of the new generation of scientists, the transmission of knowledge and skill, of insight and wisdom. The latter task is no less necessary, no less worthy. From the beginnings of human history, the exponentially accelerating growth of human power.... has required each generation to instruct and inform the next.

This is the challenge that faces every teacher of a science as he steps into the classroom or guides the early efforts of an individual student. Here, in this sea of fresh faces—here, amidst the stumblings and fumblings—may be the Newton or Einstein, the Mendel or Darwin of tomorrow. For few—so very few—men are self-taught. The teacher cannot supply the potentialities of his students, but he is needed to see that the potentialities will unfold, and unfold fully. His is not only the task of passing on the great tradition of the past, with its skills and accumulated knowledge; he must also provide breadth and perspective, self-criticism and judgment, in order that a well-balanced scientist may grow to full stature and continue the search.

Of all the resources of a nation, its greatest are its boys and girls, its young men and women. Like other material resources, these can be squandered or dissipated. They are potential greatness, but they are only potentialities. Science creates knowledge and knowledge generates power, but knowledge resides only in the minds of men who first must learn and be taught, and power is tyranny unless it be guided by insight and wisdom, justice and mercy. The greatest of men have been teachers, and the teacher is greatest among men.[15]

The Social and Ethical Responsibilities of Scientists

The scientist escapes lightly—instead of ten commandments only four: to cherish complete truthfulness; to avoid self-aggrandizement at the expense of one's fellow-scientist; to defend fearlessly the freedom of scientific inquiry and opinion; and to communicate fully one's findings through primary publication, synthesis, and instruction. Out of these grow the

[15] B. Glass, "The scientist and the science teacher," *Amer. Assoc. Univ. Professors Bull.* 50 (1964): 267–68.

social and ethical responsibilities of scientists that in the past twenty years have begun to loom ever larger in our ken.

These may be considered under the three heads of proclamation of benefits, warning of risks, and discussion of quandaries. The first of these, the advertisement of the benefits of science, seems to be sufficiently promoted in these days when science is so well supported by government and private agencies and when grants are justified on the basis of social benefits. Every bit of pure research is heralded as a step in the conquest of nuclear or thermonuclear power, space exploration, elimination of cancer and heart disease, or similar dramatic accomplishments. The ethical problem here is merely that of keeping a checkrein on the imagination and of maintaining truthfulness. But the truth itself is so staggering that it is quite enough to bemuse the public.

Since 1945 more and more scientists have become engaged in warning of the great risks to the very future of man of certain scientific developments. First the atomic bomb and then the hydrogen bomb brought swift realization of the possibility of the destruction of all civilization and even the extinction of all human life were a nuclear war to break out. The atomic scientists, conscience stricken, united to secure civilian control of nuclear energy. Albert Einstein and Bertrand Russell issued an appeal to scientists to warn the world of the tragic consequences of overoptimism and of an unbridled arms race. Joined by a dozen notable scientists, they initiated the "Pugwash" Conferences on Science and World Affairs in 1957. In these conferences scientists of East and West sat down together to talk, in objective scientific terms, of the military and political problems of the world and their resolution. It was not that the scientists at all felt themselves to be more highly qualified than diplomats and statesmen, economists or lawyers, to find solutions of the most difficult and delicate problems of international relations. They acted on two grounds only: that they understood the desperate nature of the situation about which the world must be warned in time; and that they hoped discussions by persons accustomed to argue in objective, scientific terms might pave the way for better understanding and more fruitful negotiation on the part of officials. In the ensuing discussions of the effects of fallout from nuclear weapons tests on persons now living and on the generations yet unborn, scientists played a very important role. In no small measure, I believe, historians of the future will recognize how great a part was played by the scientists in bringing about the partial nuclear weapons test ban. Scientists are now deeply involved in politics, and naturally enough often on both sides of the argument, for although they may agree upon the basic scientific facts which are relevant to the issue, there are rarely enough established facts to clinch the argument, and there is always room for differences of opinion in interpreting the facts. In such issues the relevant ethic requires the scientist to state his opinion on matters of social concern, but at the same time to distinguish clearly between what he states to be fact and what opinion he holds. Moreover, his opinion about matters within his technical sphere of competence is an "informed" opinion; his opinion about other matters, even other scientific matters, is that of a layman. He must in all honesty make clear to the public in what capacity he speaks.

Nuclear war is only one of the dire misfortunes that are poised above the head of modern man. The unrestricted and appalling rate of population increase in most countries of the world, if projected just a few decades into the future, staggers the imagination with its consequences. Effective control of the birth rate is the only conceivable answer to effective reduction by modern health measures of the death rate. This is the world problem second in importance at the present time, and it must engage the conscience of the scientist.

The problem of the future is the ethical problem of the control of man over his own biological evolution. The powers of evolution now rest in his hands. The geneticist can define the means and prognosticate the future with some accuracy. Yet here we enter the third great arena of ethical discussion, passing beyond the benefits of science and the certain risks to the nebulous realm of quandaries. Man must choose goals, and a choice of goals involves us in weighing values—even whole systems of values. The scientist cannot make the choice of goals for his people, and neither can he measure and weigh values with accuracy and objectivity. There is nonetheless an important duty he must perform, because he and he alone may see clearly enough the nature of the alternative choices, including laissez faire, which is no less a choice than any other. It is the social duty and function of the scientist in this arena of discussion to inform and to demand of the people, and of their leaders too, a discussion and consideration of all those impending problems that grow out of scientific discovery and the amplification of human power. Science is no longer—can never be again—the ivory tower of the recluse, the refuge of the asocial man. Science has found its social basis and has eagerly grasped for social support, and it has thereby acquired social responsibilities and a realization of its own fundamental ethical principles. The scientist is a man, through his science doing good and evil to other men, and receiving from them blame and praise, recrimination and money. Science is not only to know, it is to do, and in the doing it has found its soul.

Part III
Biology and Social Issues

7 Science: Endless Horizons or Golden Age?

At the end of each year, and even more customarily at the close of every decade, we are wont to look back and summarize our gains and our losses, and to look forward in the hope of forecasting as well as possible what the next year, the next decade, or even the next century may hold for us. For the scientist it seems particularly appropriate, since the future of science itself—its rate of advance in various directions and the roadblocks or morasses ahead—will likely be mirrored in the rates of technological change and consequently in the attainment of human well-being or the advent of catastrophe. One need not be a prophet to attempt an analysis and critique of scientists' views about the future of science, and these may themselves sharpen our vision of the problematical years that lie ahead.

Scientists themselves often seem little concerned about the matter. They are concerned about the quantitative measurement of all sorts of phenomena except that of the growth, stasis, or decline of their own collective activity. Young scientists, in particular, seem to find it sufficient to harbor a personal liking

---

This essay was first prepared as the address of the retiring president of the American Association for the Advancement of Science, in Chicago, at the annual AAAS meeting, on December 28, 1970. It was published in *Science* 171 (1971): 23–29 and was reprinted in a collection of the addresses of retiring presidents of the association in a volume entitled *The Maturing of American Science,* edited by Robert H. Kargon (Washington, DC: AAAS, 1974). It has been considerably revised for the present volume, and is reprinted with permission. Copyright 1971 by the AAAS.

for scientific exploration and discovery, without much worry about the magnitude of the edifice they are constructing, the depth of the mine they are excavating, or the extent of the continent they are invading. With advancing years one commences to wonder about the relative position of one's own work in a broader, more extended perspective. Thus one is led to try to evaluate present scientific knowledge not only in comparison with the past but also the future of man's understanding of nature. Then the question must be faced: Are there finite limits to scientific understanding, or are there endless horizons?

I have chosen two phrases for my title in order to represent the extremes in the spectrum of belief in the future of science: one, the view of limitlessly expanding knowledge and of infinite bounds; the other, the view that scientific knowledge, like our universe, must be finite, and that the most significant laws of nature will soon have been discovered. The Golden Age is thus the age of perfected knowledge, of consummated applications, and of social stasis. The first of these phrases is the title of a book written in 1946 by Vannevar Bush about the future of science.[1] The phrase "Golden Age" is also from the title of a book, *The Coming of the Golden Age,* written by Gunther Stent and published in 1969.[2] They thus epitomize the view of two eminent scientists, one a physical scientist and engineer and the other a geneticist and molecular biologist. Each of them was writing at the close of a period of unparalleled development in his respective area of science. In 1946, Vannevar Bush could contemplate from the vantage point of his recent work as director of the wartime Office of Scientific Research and Development what Karl T. Compton called "the greatest mobilization of scientific power in the history of the world." Not only could Vannevar Bush think with just pride of the advanced weaponry which had been created but, even more important, of the harnessing of scientific talent in new ways through operations research and systems analysis. His own inventions, the network analyzer and the differential analyzer, were harbingers of the age of computers. He foresaw the problems involved in converting atomic energy from a military

[1]Vannevar Bush, *Endless Horizons* (Washington, D.C.: Public Affairs Press, 1946).

[2]Gunther S. Stent, *The Coming of the Golden Age* (Garden City, NY: Natural History Press, 1969).

secret jealously guarded by a few nations to a worldwide basis of power production open to all nations but subject to the hazards of power politics. He recognized the difficulty in promoting scientific advance in the atmosphere of secrecy, in the face of demands for "classifying" information. He was foremost among those who planned and advocated national support of basic scientific research, which he felt had lagged during World War II, and he was one of the fathers of the National Science Foundation. It may thus have been quite natural for him to see ahead of us the "endless horizons" of scientific knowledge, the greatest challenge to moderns for whom the West had been reached, the world's continents settled and the oceans crisscrossed, for whom in fact the excitement of the geographical frontier has vanished.

Like Bush, Gunther Stent was writing at the end of the greatest advances ever made in his own scientific field. Genetics, as a science, did not even exist before the opening of the twentieth century. Molecular genetics, which Stent identifies as the third and fourth phases of the science, its so-called "Dogmatic" and "Academic" periods, began only in 1953 or thereabouts, and yet in the ensuing years it has added a greater knowledge of the chemical basis of heredity and its mechanisms than in all the years before, and seems, according to Stent, to be well into a final period of diminishing returns. More and more scientists, publishing more and more papers, fill in missing details and extrapolate knowledge in quite predictable directions. The great conceptions, the fundamental mechanisms, and the basic laws are now known. For all time to come, these have been discovered, here and now, in our own lifetime. They can never be discovered again unless man loses his scientific heritage; and even so, they cannot be discovered again for the first time. They can only be reexhumed, refined, or modified. Like Newton's laws of motion, like the Copernican solar system, like the origin of species, or like relativity and quantum mechanics, they are now the known, not the unknown—and by just so much is the extent of unknown Nature diminished.

Like Roderick Seidenburg, with whose book *Posthistoric Man*[3] Gunther Stent was apparently unacquainted, Stent expects

[3]Roderick Seidenberg, *Posthistoric Man* (Chapel Hill: University of North Carolina Press, 1950). See also Bentley Glass, "A biologic view of human history," *Scientific Monthly* 73: (1951): 363–68.

that the cessation of scientific advances will ultimately lead to an end to technological and social progress. Both Seidenberg and Stent acknowledge their debt to Henry Adams in charting the dynamics of progress and predicting its ultimate cessation. Whereas Seidenberg pictures the final state of society as post-historic, in the sense that having achieved ultimate knowledge and its applications to man's life in numerous ways, society will crystallize into a changeless pattern, Stent, foreseeing the virtual end of an exponential increase in human power in a very few centuries, and the consequent end of the arts and sciences, predicts a Golden Age, when man will live in a way "not very different from a recreation of Polynesia on a global scale." It is worth quoting his description of this final state of society:

> The will to power will not have vanished entirely, but the distribution of its intensity among individuals will have been drastically altered. At one end of this distribution will be a minority of the people whose work will keep intact the technology that sustains the multitude at a high standard of living. In the middle of the distribution will be found a type, largely unemployed, for whom the distinction between the real and the illusory will still be meaningful and whose prototype is the beatnik. He will retain an interest in the world and seek satisfaction from sensual pleasures. At the other end of the spectrum will be a type largely unemployable, for whom the boundary of the real and the imagined will have been largely dissolved, at least to the extent compatible with his physical survival. His prototype is the hippie. His interest in the world will be rather small, and he will derive his satisfaction mainly from drugs or, once this has become technologically practicable, from direct electrical inputs into his nervous system. This spectral distribution, it will be noted, bears some considerable resemblance to the Alphas, Betas, and Gammas in Aldous Huxley's *Brave New World.* However, unlike Huxley, I do not envisage this distribution to be the result of any purposive or planned breeding program, but merely a natural population heterogeneity engendered mainly by differences in childhood history. Furthermore, in contrast to the low-grade producer roles assigned to Betas and Gammas, beatniks and hippies will play no socioeconomic role other than being consumers.

> As far as culture is concerned, the Golden Age will be a period of general stasis, not unlike that envisaged by Meyer[4] for the arts. Progress will have greatly decelerated, even though activities formally analogous to the arts and sciences will continue. It is obvious that Faustian Man of the Iron Age [we of today imbued with the idea of progress] would view with considerable distaste this prospect of his affluent successors, devoting their abundance of leisure time to sensual pleasures, or what is even more repugnant to him, deriving private synthetic happiness from hallucinatory drugs. But Faustian Man had better face up to the fact that it is precisely *this* Golden Age which is the natural fruit of all his frantic efforts, and that it does no good now to wish it otherwise. Millennia of doing arts and sciences will finally transform the tragi-comedy of life into a happening.[5]

Here, then, are diametrically opposed views of the future of man which grow out of the contemplation by the scientist of the achievements of science itself. What are the basic assumptions on which these views are founded? It seems to me that the implicit assumption of greatest importance relates to the finiteness, or alternatively the infinitude, of possible knowledge. No one really questions, I believe, that even now we may be like little boys on the shores of a vast ocean, tossing pebbles into the waves, as Isaac Newton described his own feeling when confronted by the seemingly endless mysteries of nature. What remains to be learned may indeed dwarf imagination. Nevertheless, the universe itself is closed and finite, or at least bounded to our knowledge by the radius of the light-years since its beginning, about 10 billion years ago; and our telescopes have now plumbed space almost to those limits of the observable. Their growing knowledge of the universe has led scientists to believe more and more firmly that the laws of nature have universal applicability. Matter is composed of the same particles and elements everywhere. Radiant energy moves with the same speeds and has the same characteristics everywhere. Local differences are explicable in terms of local conditions and past history.

[4]L.B. Meyer, *Music, the Arts and Ideas* (Chicago: University of Chicago Press, 1967).

[5]Stent, op. cit., pp. 137–38.

The uniformity of nature and the general applicability of natural laws set limits to knowledge. If there are just 100, or 105, or 110 ways in which atoms may form, then when one has identified the full range of properties of these, singly and in combination, chemical knowledge will be complete. There is a finite number of species of plants and of animals—even of insects—upon the earth. We are as yet far from knowing all about the genetics, structure and physiology, or behavior of even a single one of them. Nevertheless, a total knowledge of all life forms is only about 2 million times the potential knowledge about any one of them. Moreover, the universality of the genetic code, the common character of proteins in different species, the generality of cellular structure and cellular reproduction, the basic similarity of energy metabolism in all species and of photosynthesis in green plants and bacteria, and the universal evolution of living forms through mutation and natural selection all lead inescapably to a single conclusion: that, although diversity may be great, the laws of life, based on similarities, are finite in number and comprehensible to us in the main even now. We are like the explorers of a great continent who have penetrated to its margins in most points of the compass and have mapped the major mountain chains and rivers. There are still innumerable details to fill in, but the endless horizons no longer exist.

It seems to me that Vannevar Bush would not really disagree with the foregoing analysis. When one examines his book, *Endless Horizons,* for the expected discussion of the challenge of scientific advancement to mankind and for some consideration of the actual meaning of the title, one seeks in vain. A great deal of attention is given to specific technological advances and to the possible further developments of the future, but the nearest to a discussion of the theme implied by the title is in two brief sections of the essay "As We May Think." These are headed, respectively, "Endless Trails" and "Horizons Unlimited." Yet the endless trails turn out to be only searches through the coded items stored in a vast mechanical memory, or "memex"; while the unlimited horizons unfold only in the form of specific predictions of various ways in which science may in the future develop the production, storage, and retrieval of information, or discover new extensions of our human senses.

Whether the horizons are really endless is an idea not only unanswered, it is not even examined. On the other hand, in his final essay, "The Builders," Vannevar Bush likens science to the quarrying of material for an edifice and the process of its construction.

> The material is exceedingly varied,... the whole effort is highly unorganized [and]... the edifice itself has a remarkable property, for its form is predestined by the laws of logic and the nature of human reasoning. It is almost as though it had once existed.... Parts of the edifice are being used while construction proceeds.... The workers sometimes proceed in erratic ways.... On the other hand there are those men of rare vision who can grasp well in advance just the block that is needed for rapid advance on a section of the edifice to be possible, who can tell by some subtle sense where it will be found, and who have an uncanny skill in cleaning away dross and bringing it surely into the light. These are master workmen.[6]

Are these not the images of one who believes in the *finite* extensibility of science? Is an edifice with a predestined form infinite in extent? It cannot be affirmed that Bush's metaphors and similes unquestionably imply the finiteness of nature and of scientific understanding, but is that not the most reasonable implication? If so, was the bold title, *Endless Horizons,* not itself a metaphor never intended to be taken literally, but supposed merely to imply that from our present viewpoint so much yet remains before us to be discovered that the horizons seem virtually endless?

If one grants that the universe itself is finite and that its laws too are universal and finite in number—or even if the universe be infinite but its nature and its laws universal—then with each new phenomenon discovered and explored, with each new law confirmed, there is an approach to the finite limits of scientific knowledge. In that case, it is less important to note the absolute bounds of knowledge at the present time than to examine the rate at which, in the past century and a half, our scientific knowledge has been expanding.

[6]Bush, op. cit., pp. 179–80.

By whatever parameters one may choose to select, the accumulation of knowledge has been accelerating exponentially during this period, with a doubling time of twelve to fifteen years. As I described more fully in *The Timely and the Timeless* (1970), the number of scientific periodicals, the number of scientists publishing or working, and the number of scientific papers have been doubling almost every ten years.[7] It is of course dangerous to equate the growth of scientific knowledge with growth in the number of scientists or their publication of papers. Yet I suppose, although I cannot demonstrate it satisfactorily, that there must be some relatively stable mathematical ratio between the total number of scientific contributions published in a given period and the number among them of highly important, significant "breakthroughs." Derek J. de Solla Price, writing in *Science Since Babylon,*[8] suggested that the "stature" of science, which I take to be equivalent to the frontier or attainment of science, doubles at a much slower rate than the increase of scientific information in the gross. He estimated that it has doubled, in the modern period, about every thirty years, the rate being maybe slower because of the cumulative nature of science, which like a pyramid or pile of stones must increase in volume as the cube of any increase in height.

I suspect that the advances in technology which are basic to progress are more directly related to the rate of this advance of the frontier, rather than to the actual total increase of scientific information. I have attempted to analyze this problem within the limits of biology for the seventeenth, eighteenth, and nineteenth centuries, and for the more limited science of genetics during the twentieth century.[9] By applying admittedly subjective criteria to select those major discoveries that were "milestones," in the sense that they were landmarks heralding a new insight into the nature of living beings and were often—though not always—revolutionary new concepts overturning older established views of how things are, I was able to demonstrate that

[7]Bentley Glass, *The Timely and the Timeless* (New York: Basic Books, 1970).

[8]Derek J. de Solla Price, *Science Since Babylon* (New Haven: Yale University Press, 1961).

[9]Bentley Glass, "Milestones and rates of growth in the development of biology," *Quarterly Review of Biology* 54 (1979): 31–53.

there was indeed a clearly exponential increase in the frequency of such advances in each half-century of biology over the three-hundred-year span from 1600 to 1900, and also within the science of genetics from 1900 to 1970 in each succeeding decade of our own century. These rates, though exponential, were considerably slower than the rates of the accumulation of all scientific knowledge. In biology, the exponential increase indicated a doubling in every fifty years; in genetics during this century, it indicated a doubling in every twenty-two and one-half years.

An interesting question then forces itself upon the mind. Does the exponential rate of increase of the truly great discoveries actually depend upon the faster exponential rate of the general increase of scientific knowledge? At least, they do seem to be strongly correlated.

Nicholas Rescher, a philosopher of science, has approached this problem in a more sophisticated analysis.[10] He reminds his readers not only of Henry Adams's Law, that the growth of science is characteristically exponential, but also of what he calls Max Planck's Law, that "with every advance [in science] the difficulty of the task is increased." This "Principle of Increasing Effort," he believes, is evident in the cost escalation of modern science, which also exhibits an exponential rate of increase. In this insuperable economic burden of pursuing more and more scientific knowledge by means of more and more sophisticated and expensive apparatus by more and more research workers, there is another, more hidden, element. Obviously, if both the total scientific output and the frequency of "first-rate" and "very important" discoveries are increasing in each field of endeavor exponentially, but the latter at a slower rate than the former, then the exponential curves will deviate more and more radically as time passes. That is to say, the "cost" of obtaining each new major breakthrough or very important discovery becomes ever greater in relation to the total volume of scientific work that must accompany it. When we reach a state of zero economic growth in industrial society, routine scientific work may be expected to diminish only slightly

[10]Nicholas Rescher, *Scientific Progress* (Pittsburgh: University of Pittsburgh Press, 1978).

as time passes, but the reduction of the "significant," the "very important," and the "first-rate" discoveries will be in that order of increasing magnitude. For the "milestones" and the "breakthroughs" the decline will be catastrophic. Only the very rare and accidental discovery that initiates an entire new science can then lead to a way out of the dilemma, and to a new period of exponential growth (as well as cost escalation), until it too reaches a plateau or declines abruptly.

Rescher himself does not believe in the philosophical limits to man's knowledge of nature. But then, as I have written elsewhere, why argue about the ultimate finiteness or potential infinity of the human knowledge of nature, if the simple consequence of exponentially increasing effort must rather soon in history lead to a logarithmic deceleration in the rate of first-rate—or even significant—scientific discoveries?[11]

It is not in fact becoming more and more difficult, as scientific knowledge grows, to make a totally new and unexpected discovery or to break through the dogmas of established scientific views? Be that as it may, the amount of extant scientific information, trivial was well as important, is currently increasing at a rate related to the absolute number of scientists, technologists, and laboratory workers in the population. I have pointed out that in the United States this rate, which produces a doubling in somewhat over ten years, cannot long be maintained, since it is much faster than the rate of population growth for the country and already constitutes about twenty percent of the entire professional labor force. I suggest that the upper limit for the proportion of "scientists and technologists" in the entire professional labor force may perhaps be about 25 percent. Further absolute gains would seem to depend upon an increase in size of the population—an increase that is already slackening—or a possible increase of the professional component in the total labor force. Here the limiting factor may well be the requirement for a modicum of intelligence and of advanced education and training on the part of every scientist or technologist.

[11] Bentley Glass, "On scientific progress and its limits," *Quarterly Review of Biology* 54 (1979): 417–19 (See p. 419).

SCIENCE AND PROGRESS

Both historian and scientist agree that progress, insofar as it can be defined and measured, must be defined in terms of man's increasing power. This was the theme of Carl Becker, one of our most notable American historians. In *Progress and Power* he humorously summed up the matter in the following words:

> All that has happened to man in 506,000 years may be symbolized by this fact—at the end of the Time-Scale he can, with ease and expedition, put his ancestors in cages: he has somehow learned the trick of having conveniently at hand and at his disposal powers not provided by his biological inheritance. *From the beginning of the Time-Scale man has increasingly implemented himself with power.* Had he not done so, he would have had no history, nor even the consciousness of not having any: at the end of the Time-Scale he would still be (if not extinct) what he was at the beginning—*Pithecanthropus erectus,* the Erect Ape-Man. Without power no progress.
>
> ...The significant fact is that the human race, so far from having any aversion from power, has at all times welcomed it as a value to be cherished. Look where we will along the Time-Scale, we see men eagerly seeking power, patiently fashioning and tenaciously grasping the instruments for exerting it, conferring honor upon those who employ it most effectively.[12]

Stent, influenced particularly by Oswald Spengler, adopts the concept of the archetype Faustian Man, whose boundless will to power locks him "in an endless strife with his world to overcome obstacles, conflict, to his mind, being the very essence of existence." Thus, while both Becker and Stent accept the dictum of J.B. Bury that the *idea* of progress is scarcely older than the seventeenth century and did not really become a prevalent social theory until the beginning of the nineteenth, they

[12]Carl Becker, *Progress and Power* (Stanford: Stanford University Press, 1936); (New York: Vintage Books, 1965), pp. 19–20.

virtually identify *all* history with the occurrence of progress through increasing power.

As long as new instruments of power are multiplied and the command over new sources of power is increased, progress may continue. But scholar and scientist alike are dubious. Said Becker:

> But it is conceivable, even probable, that the possibility of discovering and applying new sources and implements of power will in the course of time gradually diminish, or even be altogether exhausted. In that event the outward conditions of life will change less and less rapidly, will in time become sufficiently stable perhaps to be comprehended, sufficiently stable therefore for a relatively complete adjustment of ideas and habits to the relatively unchanging body of matter-of-fact knowledge of man and the outer world in which he lives. In such a stabilized and scientifically adjusted society the idea of progress would no doubt become irrelevant as progress itself became imperceptible or nonexistent.[13]

Roderick Seidenberg pushed his vision of a stable society and a completely adjusted man to the ultimate extreme. When the organization of society has proceeded to its final crystallization "in a period devoid of change, we may truly say that man will enter upon a post-historic age in which, perhaps, he will remain encased in an endless routine and sequence of events, not unlike that of the ants, the bees, and the termites.... Man may likewise find himself entombed in a perpetual round of perfectly adjusted responses.... Man will hasten along his predestined way under the illusion of attaining his freedom on even higher levels of existence, while actually sealing his fate by all the devices his dominant intelligence can command."[14] The emotions, he thinks, will have atrophied. Even consciousness, becoming no longer necessary in the absence of tension or unstable equilibrium, will evaporate.

No rate of scientific development equal to that of the past half-century can be long maintained. Henry Adams pointed

[13]Ibid., p. 98.
[14]Seidenberg, op. cit.

out that the actual doubling times are not so important as the simple existence of an exponential rate of increase, since in any case the acceleration of the process will eventually lead to staggering, and indeed impossible, magnitudes. Do we not all remember the story of the Arabian merchant who contracted with the Sultan to work for one grain of wheat the first day, two the second day, four the third day, and so doubling with each day to the sixty-fourth day corresponding to the squares on their chessboard, and how he bankrupted the kingdom? In real increases occurring for a time at exponential rates there is, as Stent has emphasized, a positive feedback. The more offspring are produced, the more adults there will be to contribute to the next generation. The fission of one bacterium into two provides the basis for the next doubling. The recognition of progress in the economic, social, and political condition of man led to the expectation, beginning at about the end of the eighteenth century, that progress not only was possible but was to be expected, and this expectation led men to push for further progress.

"Evidently," Stent says, "the element of positive feedback embodied by progress is that the rate at which man can gain more power over the outer world is the greater the more power is already at his disposal." Yet, the argument continues, "this very aspect of *positive* feedback of progress responsible for its continuous acceleration embodies in it an element of temporal self-limitation.... If one examines one by one the parameters conceivably relevant to estimating the rate of progress, such as world population and energy consumption, per capita income, or speed of travel, one must conclude that none of them is likely ever to exceed some definite bound."[15] The further we have advanced on the accelerating curve, the closer we inevitably come to the time when limiting factors will curb the growth and bring it into some degree of stasis, or equilibrium, or possibly decline.

We may find reason to disagree with various details of these pessimistic visions of man's future, but can we honestly set aside the conclusion that *progress,* in the sense of ever-growing power over the environment, must soon come to an end? We may for a long time to come retain our emotional

[15]Stent, op. cit., pp. 92–93.

responses, which to so great an extent depend upon the hormonal balance stabilized in the early phases of mammalian evolution, long before man was man. I do not accept Seidenberg's supposition that in the keen battles between reactions on the basis of intelligence and reactions on the basis of emotion, intelligence will win the upper hand. Seidenberg seems to have mistaken the basis of emotion with instinct. True, in vertebrates such as birds or even mammals, instincts appear often to operate through the machinery of the hormones, but in bees, ants, and termites instincts depend fully as much upon the inborn connections of neurons, in other words, upon the same fundamental mechanisms as does intelligence. The distinction to be drawn is between the inborn patterns of the nervous system, which develop irrespective of the external situation, most often in vertebrates before birth or hatching from the egg, and the malleable patterns which can in some way be modified through experience and learning. Stent certainly has not envisaged a decline of emotionality—quite the contrary. His "beatniks" and "hippies," who will comprise the majority of future human populations if he is right, will react to life even more than now on the basis of emotionality.

That is a side issue. Let us keep clearly in mind that the primary basis of agreement of Becker, Seidenberg, Stent, Rescher, and perhaps even of Bush, is that progress cannot continue indefinitely. Indeed, so awesome is already the accelerating rate of our scientific and technological advance that simple extrapolation of the exponential curves shows unmistakably that we have at most a generation or two before progress must slacken, whether because the world's population becomes insufferably dense, or because we exhaust the possible sources of physical energy or deplete some irreplaceable resource, or because, most likely of all, we pollute our environment to toxic, irremediable limits. Many scientists have in recent decades examined these processes and have tried to flag the runaway express. The present more general outcry against unlimited population growth and heedless pollution of the environment offers a slight ground for hope. The prime difficulty is that so many persons, not only in the highly industrialized countries but even more among the peoples of the underdeveloped countries, now see their hopes for the future bound up with the continuation and extension to all mankind of the progress hitherto limited to a few fortunate lands and people. The momentum of these processes is furthermore such that measures to apply the brakes to population growth or to reclaim and preserve our environment, even if the measures are firm and effective, will take at least one generation—say until the year 2000—to reach full effect.

To say the least, capitalism has not yet found a way to survive on the basis of a simple replacement economy. The pressure to keep the gross national product growing—in real, not inflationary, terms—comes from all elements of our society, from labor as well as employer and industrialist, from natural scientist as well as economist. (True, a few rumblings of dissent from this point of view have been voiced.) Dependence upon the profit motive, which has in the past stimulated the unparalleled growth of Western Europe, Japan, and the United States, just as surely presses society into overexploitation.

Socialism, which, as it now operates in other segments of the world, is but a modified governmental capitalism, seems no better able to avoid overexploitation than does pure individualistic capitalism. No socialist country has met the population problem as successfully as highly industrialized, capitalist Japan. The depletion of the world's resources flowing from economic imperialism is as much a characteristic of socialism as of capitalism. The pollution of the environment proceeds with perhaps even fewer checks in a socialist land than in a Western democracy, where the outcry of protesters is freer and the influence upon legislators more direct.

Humanity indeed faces hard times unless a new social and economic system, far more responsive to human needs and far more foresighted than in the past, can be invented. What is coming to be called technology assessment is one example of what is needed: a complete systems analysis of all the long-range effects and side effects of each technological alteration, before it is unleashed. Our local, state, and national governments are hopelessly unprepared to exercise such a function. Our legislators and bureaucrats are uninformed and unaware, for the most part, of the crisis facing the world today, and they are perhaps universally incapable of conducting the kind of analysis required. What they could do, if they would, would be to establish—and fund—in sufficient number and strength the agencies to carry out the studies required; and they could unquestionably, if they would, assess strict penalties upon all, high or low, who waste our resources and needlessly pollute our environment.

In Czechoslovakia, in 1966, I discovered the existence of an Institute for the Study of the Biological Landscape.[16] It was a response to an overwhelming technological disaster that occurred when, during the industrialization of Slovakia after World War II, a great mill was constructed in a narrow valley of the Tatra Mountains where an atmospheric inversion layer penned down the fumes, heavy in sulfur dioxide, arising from the combustion of local coal. Every living thing in the valley had been killed, and the workers, so I was told, at times had to wear gas masks and live many miles away, at a great loss of time. In the United States—and in every other country, whether heavily industrial or newly entering the industrial phase—we need a profusion of such agencies for the study not only of the environment and its alteration by technological innovation, but also of the systems for predicting and regulating change. I suggest, at the risk of grave misunderstanding, that in future histories of the world the decade of the 1960s may be known most significantly as the time when man, with unbridled lust for power over nature and for a so-called high standard of living measured by the consumption of the products of an industrial civilization, set in motion the final speedy, inexorable rush toward the end of progress.

I am not a hopeless pessimist, however. I have already suggested one way in which mankind might avoid the debacle. These agencies to curb irrational exploitation, widespread in all countries, could mutually reinforce one another. They might form the basis of a most effective United Nations network. In any case, what the United States itself most needs in this juncture is a second Vannevar Bush who could organize and direct a comparable national effort in peacetime, as Bush did in time of war. It is consequently disheartening to observe that the technological developments most needed for the kind of systems

[16]Bentley Glass, "For full technological assessment," *Science* 165 (1969): 755; "Biological aspects of technology assessment," *Quarterly Review of Biology* 45 (1970): 168–72.

analysis required, and which Bush foresaw in 1946, are not yet available. His great "memex" system, which would replace private files and public libraries, which would code all extant information on every subject for instant retrieval, and which would provide endless cross trails of reference to stimulate fresh experimentation and analysis, is still unrealized. There seems to be little or no reason why such a system could not be constructed today. The computers, now far exceeding in capacity for storage and rapid retrieval anything Vannevar Bush had dreamed of, are here. What is still lacking is the compatibility of different systems and the actual realization of the conception. The machines, in fact, are far superior to the skill of the human programmer who feeds our mechanical memory banks. In consequence, the very accumulation of disorganized scientific information is burying us daily deeper in a profusion of unevaluated, unused knowledge. Scientific information has increased a hundredfold since 1900, but most of it is already forgotten. The next age of scholarship will no doubt promote processors and analysts who need only delve in the mountains of extant scientific and technological literature for forgotten and uncomprehended items of knowledge.

THE MAN OF THE FUTURE

There is one other way in which the crisis provoked by an uncontrolled exponential increase can be surmounted. Every biologist who deals with the population growth of bacteria, mice, or men knows that exponential growth not only must come to a halt when one or another factor of the environment becomes limiting, but also that subsequently a variety of fates may ensue. The population may die out altogether if it cannot adjust rapidly enough. Toxicity may increase, for example, owing to the wastes of the population itself, to such an extent that even in the presence of ample food, all organisms perish. Or fish, in a polluted lake deprived of sufficient oxygen, may die by millions. On the other hand, the population may level off toward a condition of stability, in which proliferation and exploitation of the environment are balanced against the capacity of the environment, especially the food supply, to regenerate. Or, perhaps most interesting of all, after a period of such stabilization the once-limiting factor may become no longer the limiting parameter of the environment. Then the population may set out on a new cycle of exponential growth.

If man learns in time to regulate his numbers so as not to pollute and exhaust his environment irreparably, if he learns to utilize ordinary rock in place of rare metals, or to synthesize out of abundant plant materials most of what he needs, it is possible that the limiting factor may become the present supplies and sources of energy. Fossil fuels are exhaustible, perhaps within a century; nuclear energy from fission depends also upon scarce and limited amounts of uranium or artificial transuranium elements. Man's future resources of energy may thus depend upon controlled nuclear fusion, or alternatively upon the use as fuels of hydrogen released from water by solar energy. Then we might indeed see a new phase of exponential growth of human *power* (though not necessarily of population).

In the new, far more regulated society of man which will inevitably be forced upon us by our exponential rates of increase, the present genetic types of mankind may not all permit a happy adjustment. The nature and personality of man must change, too, if progress is no longer to be our chief goal and ambition. The once-sacred rights of man must alter in many ways. Thus, in an overpopulated world it can no longer be affirmed that the right of the man and the woman to reproduce as they see fit is inviolate. On the contrary, if my own additional child deprives someone else of the privilege of parenthood, I must voluntarily refrain or be compelled to do so. In a world where each pair must be limited, on the average, to two offspring and no more, the right that must become paramount is not the right to procreate, but rather the right of every child to be born with a sound physical and mental constitution, based on a sound genotype. No parents will in that future time have a right to burden society with a malformed or mentally incompetent child. Just as every child must have the right to full educational opportunity and a sound nutrition, so every child has the inalienable right to a sound heritage.

The foregoing remarks may be subject to a grave misunderstanding. In a totalitarian society, it is true, one would expect that the leaders and the bureaucracy would attempt to enforce limitations on parenthood and to regulate, with stern penalties for disobedience, the eugenic control of the population. In a democratic society, the values of which I cherish, similar constraints must be voluntary. We already see that, in the United States as well as in some other countries, voluntary measures are sufficient to reduce the average size of family to less than two children per couple. We may similarly expect that, when sufficient genetic information respecting the availability of prenatal diagnosis and selective abortion is quite general, and when clinics abound in which couples can receive sympathetic but accurate advice about the prospective well-being of their children, the great majority of parents will take advantage of the opportunities. Genetic counseling, of a sort not common today, counseling that takes into account the many varieties of factors, social and economic, psychological and personal, as well as healthy well-being and promise for a sound physical and mental development of the child, will need to become merged with marriage counseling in general. When that day arrives, we can confidently expect that *voluntary, informed choice* will achieve what dictators and tyrants vainly seek by force—the improvement of the prospects for life of each child born into the population.

Human power is advancing with extraordinary rapidity in this realm of control over the genetic characteristics of the unborn. Perhaps, as Carl Becker so pregnantly stated, our race, far from having any aversion from power, will welcome this power too, will seek it, fashion it, and grasp it tenaciously. Unlimited access to state-regulated abortion will combine with the now perfected techniques of determining chromosome abnormalities in the developing fetus to rid us of the several percentages of all births that today represent uncontrollable defect—such as mongolism (Down's syndrome) and sex chromosome abnormalities such as the XYY type. Genetic clinics will be constructed in which, before long, as many as 100 different recessive hereditary defects can be detected in the carriers, who may then be warned against (or prohibited from) having offspring. Continuing efforts to synthesize genes and to introduce a sound gene by means of a carrier virus into a child or a fetus bearing only the defective allele are promising. They may make it possible not merely, as now, to correct a genetic defect while leaving intact in the individual the defective gene to be passed on, but instead actually to substitute a sound gene for the defective gene in the reproductive cells of the treated person. This procedure will unquestionably be most effective if carried out during the early embryonic stages of development, or in the just-fertilized egg itself, before it has begun its cleavage into numerous cells. Hence we must look with expectant attention at the startling progress that has been made during the past two decades in the laboratory of R.G. Edwards at Cambridge University, England, and increasingly in other places.

Dr. Edwards and his collaborators[17] have succeeded in obtaining considerable numbers of oocytes from human females, culturing them in media that permit their maturation, fertilizing them with fresh spermatozoa, and observing normal development of the embryos so produced to the blastocyst (hollow ball) stage at which they normally become implanted in the wall of the mother's uterus. The world followed with absorbed attention the results of the first successful implantation and rejoiced with the parents at the birth of a fine baby, little Louise Brown. Since that time several other successful embryo implants have been made, not only in England but also in Australia, the United States, and, reportedly, in India.

The way is thus open to what I have called "prenatal adoption," for not only might the selected embryos be implanted in the uterus of the woman who supplied the oocytes, but in that of any woman at the appropriate time of her menstrual cycle. Edwards would cautiously limit the application of his developing techniques to the provision of a healthy embryo for a woman whose oviducts are blocked and therefore prevent descent of the egg. It should be obvious, however, that the technique can be widely extended. The embryos produced in the laboratory might come from selected genotypes, both male and female. Preservation of spermatozoa in deep-frozen condition could permit a high degree of selectivity among the sperm donors, who so far have been limited to the husbands of the women donors of the oocytes. In 1981 it was reported that ovarian culture was successfully carried out, with ovulation of

[17]R.G. Edwards and R.E. Fowler, "Human embryos in the laboratory," *Scientific American* 223 (December 1970): 44–54.

mature eggs ready for fertilization. We may therefore expect that ovum banks will in the future take their place alongside the existing sperm banks, and that embryos for implantation will be available from select donors of both sexes.

The successful freezing and storage of embryos, and their subsequent thawing and implantation in the uterus of a female, the embryo's own mother or a surrogate, has long been practiced successfully in animal breeding. It has now been extended to human childbirth, for a dateline from Melbourne, Australia, of April 10, 1984, carried by the *New York Times* of that date, reports the successful delivery of a baby girl produced from a frozen embryo. The frozen storage for two months was required to bring the embryo's implantation to the most probably successful period of the mother's reproductive physiology. There is no reason, however, why much longer terms of frozen storage, and implantation in a surrogate mother in place of the natural one, could not be carried out successfully.

Many novel questions of good judgment and of ethics will have to be faced at that juncture. Again, I believe, voluntary, informed choice will be preferable to legal restrictions on the options that will be available, as some persons would advocate, or preferable, on the other hand, to active governmental involvement in the production of each new generation of human beings.

Sex determination of the embryos is possible before implantation; and embryos with severely abnormal chromosome constitutions can be discarded. By checking the sperm and egg donors with a battery of biochemical tests, mating of carriers of the same defective gene can be avoided, or the defective embryos can themselves be detected and discarded. By preserving the reproductive cells obtained from young persons under conditions which minimize mutation, those same individuals may even have their own offspring at a relatively advanced age without incurring the higher probability of adverse gene and chromosome defects such as normally increases with age of the parent. In the future of mankind it will become possible for every person to procreate with assurance that the child, either one's own or one prenatally adopted, has a sound heritage capable of fully utilizing the opportunities provided by society for optimal development.

We do not know enough about the more complex aspects of human nature, such as intelligence and personality, to exercise strong selection wisely. However, were every couple to have only two children, or be permitted to exceed that number only upon special evidence that the first two are physically and mentally sound, a mild eugenic practice might be introduced that is probably all mankind is prepared to accept at the present time.

Let us insist, in any case, upon avoiding any measures that might decrease the extent of human genetic diversity. The intelligence of man is an evolutionary product of natural selection for adaptability to great variation of surroundings to tremendous vicissitudes of experience, as Henri Bergson concluded seventy years ago. Such evolved intelligence depends upon a genetic pool in human populations to which many genes contribute, in ever-shifting combinations produced by the patterns of heredity and reproduction. A pure line, no matter how perfectly selected for special excellence under given conditions, could be well adapted only to particular conditions. Unless man's power produced a world not only stable but unvarying and unvaried, the maintenance of human diversity should remain a paramount aim. As long as our brave new world presents an abundance of choices and as long as we have freedom to choose, just so long will human intelligence based upon genetic diversity continue to be a primary requirement.

Intelligence without integrity will fail. Roderick Seidenberg, in a later book, *Anatomy of the Future,*[18] put it well. "The means rather than the ends of life are in our command." Man requires a challenge and a quest if he is to avoid boredom. The Golden Age toward which we move will soon look tawdry if we no longer see endless horizons about us. We must, then, seek a change within man himself. As man acquires more fully the power to control his own genotype and to direct the course of his own evolution, he must produce a Man who can transcend his present nature. The Erect Ape-Man had no conception of the power that his twentieth-century descendant would wield in really so short a span of evolutionary time. For the Erect

[18]Roderick Seidenberg, *Anatomy of the Future* (Chapel Hill: University of North Carolina Press, 1961), p. 26.

Ape-Man a club was the acme of power. Even so, if Man can avoid the ultimate follies which our present powers have bestowed upon us and can survive a few centuries more, we today can little perceive what mankind may become. Perhaps indeed the Golden Age of no progress will be but a passing phase, and history, recording new changes, may resume.

8 Education in Our Changing Time

Too often we forget that a "liberal" education is basically one that frees the mind and extends the scope of the individual's appreciation of human knowledge. A liberal education is not a matter of choosing this or that curriculum; it is a progressive freeing of the mind. From the dawn of history, education has had two primary functions, the one being that of transmitting from each generation to the next the knowledge requisite to human power and of simultaneously extending it, the other being that of enlarging our comprehension of humanity's place in the universe. The first of these is the technical aspect of education; the second, the philosophical. Both are necessary. "Man must eat, but man cannot live by bread alone."

Neither of these aspects of education was at the beginning, in a true sense, *liberal.* The first was empirical. It grew slowly and painfully out of trial and error into arts and crafts and ultimately into a scientific technology. The other began in magic, superstition, and primitive religion, and evolved in blood and tears into the cosmology of Copernicus and Galileo, the evolutionary theory of Darwin, and the psychology of Pavlov and Freud. Only in the last 350 years have these developments of

---

This essay restates in part thoughts I have previously developed in "Liberal Education in a Scientific Age" in *Science and Liberal Education* (Baton Rouge: Louisiana State University Press, 1959); in one of the Spaulding Distinguished Lectures at the University of New Hampshire in 1966, "Education in Our Changing Time," (Durham: University of New Hampshire Press, 1968); or in "The Obsolescence of Education" in *The Timely and the Timeless* (New York: Basic Books, 1970). Unattributed quotations are from those essays. Copyright permission granted by the University of New Hampshire.

human thought and education become fused in the growth of modern science. The functions of science in society thus parallel those of education. On the one hand, science is concerned with the development of increasingly adequate concepts about man and man's place in the universe. On the other hand, science has acquired the function of enlarging man's command over nature (including his own nature) by means of new knowledge acquired through systematic, accurate observation and controlled experimentation. The freedom of the mind is enlarged in the first case through liberation from the shackles of ignorance and superstition. It is enhanced in the second case through the enlargement of choice and opportunity. To paraphrase Lyman Bryson: Man is not free to choose what he has never heard of or what doesn't exist.[1]

I have therefore claimed on occasion that science is the greatest liberating and liberalizing force in human thought. It must, if this be true, be made an integral, central part of the liberal education. It is not peripheral in the education of modern man, who so desperately seeks adjustment to his frenetic civilized environment. It is crucial—the core within the apple, the skeleton within the body—for it carries the seeds of change and progress in civilization, and it is the basis of support, giving form and shape to the social sciences, arts, and humanities of our time.

The great problem of education is to integrate the teaching of the sciences with the rest of the liberal studies. As for the sciences, it is far more important how they are taught than how much time is devoted to them or what particular content is emphasized. The teaching of the sciences in liberal education must above all else reveal its nature as an instrument whereby man approaches objective truth about the material world and the living beings that inhabit it. For this, as our great American historian Carl Becker so unforgettably emphasized, is the basis of human progress and the source of human power over all forces without and within.[2] Hardly second in importance is the need to reveal the social and humanistic relations of the sciences.

[1]Lyman Bryson, *Science and Freedom* (New York: Columbia University Press, 1947), p. 5.

[2]Carl Becker, *Progress and Power* (Stanford: Stanford University Press, 1936).

As for the social studies, the arts, and the humanities, it is equally necessary to transform their teaching so as to make clear their interdependence with the sciences. I dare assert that history is no true history without some revelation of the role of the sciences and of technology in the ascent of man to his present state of power. The social sciences are in fact no sciences without a grasp of the economic, political, and social problems that grow out of technological change based on scientific advances. The critical problems of mankind today—the control of nuclear power and the dangers of nuclear war, the population explosion, problems of health and the lengthening life span, and the pollution and destruction of the human environment—and the even more crucial problems of tomorrow—arise from the powers vouchsafed by science, and their solutions require scientific as well as social understanding.

Communication in all languages is a problem of the sciences and a scientific problem no less than a linguistic and literary one. The international scope of science and the need of a universal language for scientific communication press powerfully upon all language. Douglas Bush has said that "All modern poetry has been conditioned by science, even those areas that seem farthest removed from it."[3] Music, surely, is audible mathematics. The fine arts spring from nature, and the study of nature is basic to art. To be sure, the discernment of order in nature and in the formulations of mathematics does not circumscribe the sense of beauty. Truth may be beauty, but beauty need not be truth. Nevertheless, if art forsakes nature altogether, it not only abandons the function of keeping men "continuously aware of the freshness and ineffable beauty and richness of the immediately apprehended," as F.S.C. Northrop defined its primary function, but it furthermore vitiates its second function, that of supporting some doctrine believed to be true. For art has within itself, he says, no criterion of truth, but for that must rely upon science and philosophy.[4]

Every man, if he is to avoid confusion of spirit, must create for himself an integrated view of himself and his world. This is

[3]Douglas Bush, *Science and English Poetry* (New York: Oxford University Press, 1950), p. 151.

[4]F.S.C. Northrop, *The Logic of the Sciences and the Humanities* (New York: Macmillan, 1948), p. 169.

the function of philosophy, and in carrying out that function it must work within the framework of scientific knowledge and concepts. We sadly need today a philosophy that will embrace the rights and the needs of the individual and also the welfare of society, as the Judaeo-Christian outlook so successfully has done for many centuries. At the same time, however, the new must be a dynamic philosophy, changing with the basic shifts in man's scientific understanding of himself and his world, working toward a higher synthesis and not irrevocably tied to Newton or Einstein, Pavlov or Freud, Darwin or Mendel. For all of these, however great, have seen through a glass darkly. The whole of nature is far beyond man's present comprehension, the edifice of science and philosophy a mere foundation and not the complete structure it will some day be. We indeed hope to build of our ideas and conceptions a cathedral, vast and beautiful, time-tested, wherein the human spirit may find strength and courage, peace and wisdom.

For all of this, science has limitations as well as powers. It tells us much, but hardly everything. It can deal with matter and energy, space and form and time. It scarcely measures values; it is thwarted by intangibles. Science reveals truths, but perhaps never the whole of truth. Its grandest conceptual schemes and theories may fail and have to be replaced. It is objective, not subjective, and the inner life of man is, and must always remain, subjective. Science is the product of the human mind, but what the mind is we do not know.

Science, as Bertrand Russell said, can enhance among men two great evils, war and tyranny.[5] For the powers of science can be used for evil as well as good ends. How ironical that science, the greatest force in liberating the mind, might be turned to enslave and destroy us! Yet this is true of all power. As mankind, with previously unimagined physical and biological powers, stands lonely and afraid in the universe, it is of ourselves that we are afraid, of the choices that must be made between good and evil.

What I have said clarifies, I hope, the interdependence of the sciences and the humanistic studies in the truly liberal education. Science, in ever greater measure, must permeate the

[5]Bertrand Russell, *The Impact of Science on Society* (New York: Columbia University Press, 1951), p. 51.

study and teaching of each of the arts, humanities, and social studies. These, on their side, must mollify, enrich, and protect the sciences.

Our world is in explosive change. The human condition in the year 1980 differed from that of 1920 more widely than the human condition in 1920 differed from that at the beginning of our era (A.D. 1). Above all else, the education of young people must be related to such facts, for they must adjust themselves to further incredible changes in human society. Traditional occupations will disappear, some almost overnight; new occupations will emerge. Other will be transformed beyond recognition. Medicine, for example, which until now has been concerned mainly with the healing of disease, will perforce concentrate more on the correction of inborn defects and on the prevention of illness through community health care—especially on the care of the aged, who will make up a rapidly increasing part of the population.

A factor of overwhelming importance in respect to biology and the other sciences is their exponential rate of growth. At least in this century and in the more rapidly developing sciences, the fund of significant knowledge is doubling every ten to fifteen years. This fact is hard to appreciate, but it follows from the rapid growth in number in the United States of the profession composed of scientists and their technicians. According to the United States Census, this has for a number of decades been the most rapidly increasing of all professions, and in absolute numbers also it is close to the top. It is surprising, perhaps, but readily calculated, that the number of scientists now alive and publishing their new findings is greater by far than those of all previous times taken together. In fact, scientists now living constitute about 90 percent of all who have ever lived. With a doubling of scientific knowledge in every decade, the amount at the end of a person's life will be about one hundred times what it was at his life's beginning. A college graduate of 1960 who then had a good grasp of the main ideas of the life sciences will scarcely know half of those current in 1970 and a fourth of those regarded as important in 1980, even if forgetting nothing. The flood of recorded knowledge spread through some 20,000 biological journals publishing half a million original researches annually is like that from a bursting dam. The ablest biologists struggle in vain to cope with it.

Let me illustrate briefly from my own special field of science, genetics. It was only in 1946 that the first critical evidence was produced to indicate that the chemical basis of inheritance is not, as had been supposed by everyone until that time, to be sought in protein but rather in a previously neglected chemical substance—one sort of nucleic acid known as deoxyribose nucleic acid, commonly abbreviated DNA. Even five years later, by no means all geneticists had been convinced that the DNA molecule contains sufficient complexity to constitute all the thousands of kinds of genes in each of the millions of species of plants, animals, and microbes. Yet by 1953 the evidence was overwhelming, and in that year the structure of DNA was formulated in the light of new theory and new methods. Shortly thereafter geneticists and biochemists began to talk of the "genetic code," the sequence of purine and pyrimidine bases in the DNA molecule that transfers the hereditary instructions to another kind of nucleic acid molecule, ribose nucleic acid (RNA), which in turn is the messenger that determines the sequence of the twenty sorts of common amino acids that become inserted in place during the synthesis of each sort of enzyme or structural protein molecule in the living cell. By 1956, just ten years after the opening development in this scientific drama, the first crude prototypes of natural DNA and RNA molecules were being synthesized in the test tube; yet still no one could guess how the code might be deciphered or whether it was the same or different in the multitudes of living species. By 1961 new methods were invented, and the code began to be deciphered. By 1966 it was completely known, and its universality among living beings was clear. Thus, in twenty brief years, an entirely new basis was achieved for understanding heredity and development in man and mouse, tree and flower, whale, bacterium, and virus—an achievement of our time second in importance and ultimate consequences only to the control of nuclear energy, if second even to that.

Later, in the 1970s, genetic engineering was born. Devices whereby bacteria and viruses transfer genes from one organism and one species to another were discovered. Enzymes—restriction enzymes, as they are called—were found to exist in abundance and in all species. These enzymes cut the DNA into pieces at particular joints, so that by using appropriate combinations of such enzymes pieces of DNA corresponding to single genes can be liberated and isolated. These can be identified by so-called hybridization in the test tube with known standard pieces of natural or artificial DNA; and then they can be introduced into viruses or bacterial cells where they become actually incorporated into the total genotype of the organism. Because all species of organisms utilize the same genetic code and possess the same sort of protein-synthesizing machinery, it follows that the transfer of human genes for producing insulin or interferon or growth hormone into colon bacilli makes it possible to produce abundantly and cheaply, within the bacteria, the once exceedingly scarce and expensive compounds needed for specific treatment of genetic disorders. To extract these products, to separate them from possible toxic substances in the bacteria, and then to use them in medical treatment requires great care but is not too difficult. The fear that transformed bacteria might in some way become monstrous agents of new disease is mitigated by the use for the engineered process of only a strain of bacterium that carries a double mutation preventing it from living at normal human body temperatures. What such an advance of knowledge has done to all the fine biology textbooks and curricula of twenty, or ten, or even five years ago can readily be imagined!

No exponential rate of increase in anything at all can continue indefinitely. Perhaps already the increase of scientific knowledge is beginning to slacken. Yet there are no clear symptoms of our arrival at any such plateau. On the other hand, the wide recognition that today national military power and security, as well as economic welfare and the standard of living, all depend upon the wider use of scientific technology is forcing an ever greater governmental and industrial support of science, although that is mainly a support of applied science, not of basic research. Sooner or later, in this century or the next, diminishing returns will slow down the exponential growth of science to a modest pace, but just at present there is no sign of a slackening.

The science textbook is today out of date and in need of revision in about five years. The knowledge of a science teacher undergoes a similar obsolescence. Unless he keeps learning at a pace that demands ever greater effort, his stock in trade depreciates at about the same rate as an automobile; in five to eight years, unless replenished and renewed, it is practically worthless.

The greatest challenge which education must face in our time is that of coping with the rapidity of change in science, in technology, in human power, in the conditions of man's life. The crux of adult continuing education lies precisely here: that the educated man of yesterday is the maladjusted, uneducated man of today and the cultural illiterate of tomorrow. Education must clearly become a continuing process of renewal.

The obsolescence of education differs only in degree in the several sciences and even in the humanities. I have heard a distinguished historian, who had been in university administration for a mere five years, say that his field had advanced so much in the interim that it would take him a year of concentrated study just to catch up with its development. In spite of a very general recognition of these hard facts, little is done to alter our pattern of education to cope with them. Physicians continue to practice medicine although fifty years have elapsed since their youthful preparation. Dentists do the same. Lawyers and engineers live comfortably on their antiquated stock in trade. Teachers slide steadily downhill in their grasp of new developments in their own subjects. What is perhaps worse than any of these is the virtually universal ignorance, on the part of educated men and women, of any advancement of knowledge outside their own professional specialties. Surely, we need a complete and thoroughgoing change in attitude toward "adult education," a careful planning of appropriate programs and courses designed for the intelligent adult out of touch with his new world, and a mandatory, cyclic renewal of training for the professional specialist.

What can be done? I believe that the Japanese have pointed the way.[6] During a visit of several weeks in the summer of 1965, I was privileged to observe the work of six of their Science Education Centers, of which there are now more than thirty, located in nearly every prefecture and major city. The first centers were founded, remarkably enough, in a country with a strong centralized Ministry of Education, entirely through local initiative in various school districts in which a few farsighted educational leaders had come to realize that drastic and novel steps must be undertaken to reduce the obsolescence of trained

[6]For a fuller account of the Japanese Science Education Centers, see Bentley Glass, "The Japanese science education centers," *Science* 154 (1966): 221–28.

science teachers in the schools. The organization of a typical science education center is well worth describing. In a building erected by the local Board of Education as a part of the school system, one finds usually four laboratories accommodating classes of about twenty-four students, for physics, chemistry, biology, and earth sciences respectively. Adjacent to each laboratory there is a large room of similar size for the staff of each department and for storage of equipment, preparation of materials, and so forth. In addition, each building is likely to contain a modest library, perhaps a lecture room for up to 100 persons, offices for the administration and in some cases a dining hall and dormitory rooms for teachers in residence. Besides equipment very like that of a typical Japanese high school, the center may have collections of minerals and fossils, a greenhouse, a small animal collection and pieces of expensive equipment such as a donated electron microscope, an X-ray machine, or a 15 cm telescope, which the teachers can learn to use and demonstrate and to which they can later bring their own classes for study and observation.

The most remarkable aspect of the centers, however, lies in their staffing and programming. Each center has a permanent staff, generally composed of a scientist and one or more highly experienced high school teachers for each discipline. These staff members plan and prepare, as well as teach, the courses of the center. There are two kinds of courses, for the most part. Short courses, five days long, are given in intensive fashion, utilizing the best of new materials and methods. I was surprised to find that the staff members were very well acquainted with our American science curriculum studies and were making full and effective, indeed very critical, use of them. The majority of teachers in the local school system, embracing often upwards of sixty schools, are thus provided with effective stimulation and renewal of training. Because teachers in Japan are employed on a twelve-month annual basis, they are subject to duty the year round, and the science education centers are never empty. During the school term, to the maximum extent possible, teachers are released from regular duties to attend the week-long course while their places are taken by substitutes or other arrangements are made for the students by flexible scheduling. The second type of course is one of great scope, occupying a group of twenty-five teachers a full half-year on leave from teaching duties. The depth of training possible in a course that runs all day, five days a week for a half-year is indeed excellent and explains the fact that I saw here the best demonstration teaching with high school students I have ever observed. The product of such courses is a group of master teachers who are dispersed to different schools and who there effectively revitalize the teaching in their subjects.

The effects of this relatively recent educational experiment were so dramatic that within a year or two after their creation the Ministry of Education stepped in to support the operating budgets of the centers up to about one-third of the annual cost and also to aid in the establishment of new centers. By 1965 the results were so extraordinary that it was decided without further trial to extend the centers to include mathematics, social studies, language, and other disciplines.

That visit to Japan was almost twenty years ago. You may well expect that conditions have changed. I therefore refer to an article by Fred M. Hechinger, education editor of the *New York Times,* in which he deplores the loss of all the gains made in the early 1960s in science and mathematics education through the support made possible by the National Defense Education Act of 1958 and the curriculum studies and institutes for science teachers funded by the National Science Foundation.[7] He sadly speaks of those years as "the golden age of science education." He then quotes Professor Michael W. Kirst of Stanford University, who has made a recent visit to Japan. Kirst has reported that "in the 1980s the Japanese education system is much better equipped than its U.S. counterpart to produce workers with a high level of skill in math, science and engineering that the economy of the future will require." Kirst noted that teaching materials developed in the United States, in particular by the curriculum studies supported by the National Science Foundation, are widely used in the Japanese schools. Above all, he emphasized that the key to Japan's quick introduction of the latest scientific developments into the schools is "its enormous investment in keeping science teachers up to date." That is

[7]Fred M. Hechinger, "The Science Age Is Turning Darker," *New York Times,* September 22, 1981, p. C6.

done, of course, by means of the science centers and their institutes, which touch virtually every science and mathematics teacher in the country.

In the United States, where local and state boards of education are so fearful of federal intervention and domination, the pattern established by the Japanese Science Education Centers offers particular encouragement. Why cannot our universities, which have until now ventured so timorously into the vital task of continuing education, cooperate with local school systems to establish regional education centers that are more than hotels or conference centers—that instead would answer to this most serious challenge to our social fabric? Federal assistance is all very well, but it is quite evident that our summer institutes, even though now in existence for several decades, skim off only the cream of the teaching profession and leave the majority of the teachers unaided—in fact, more obsolete in knowledge of their subjects and of improved ways of teaching with every passing year. An even more serious indictment of the existing summer institutes and continuing education programs is that, with the rarest exceptions, no serious continuous thought and effort have been given to design courses that are really suited to the needs of a teacher afflicted by educational obsolescence. Is it not self-evident that the kind of course well suited for a freshman or sophomore college student is not at all suitable for a teacher of the subject, a person who has already devoted much study to his or her subject but is now in need of adequate explanation of significant new developments and their relationships to the older materials? Nor is the usual course designed for a junior or senior college student, much less for a graduate student, at all useful for a renewal of training and an extension of knowledge for the practicing teacher.

The curriculum studies of the 1960s in the United States at the time revitalized the teaching in the secondary schools of physics, mathematics, chemistry, biology, and earth sciences. Later they were extended to include elementary school science and mathematics. The greatest aim of those science curriculum studies of the 1960s was to make it possible for young students to explore scientific inquiries themselves, to learn how science is done and to acquaint themselves with its methods, rather than to stuff themselves full of accumulated factual knowledge.

I had the pleasure myself of participating in one of those curriculum studies. To have contributed to such a program and to have seen it bear fruit in schools across the country has been one of the greatest joys of my life. And yet, sadly, one must recognize that very few of the teachers now in our schools are prepared to take new curricula of that sort and utilize them well. Often their enthusiasm outruns their knowledge and current level of skill. Even so, the new curricula, though no more than a step in the right direction, must be continuously revised to keep pace with the advances of science and the development of new human problems. Teachers of the sciences must be trained not once only, but many times over, throughout their professional lives.

One of the most ominous of present policies espoused by President Reagan was his decision to eliminate the educational programs of the National Science Foundation (NSF), which in the past were the chief support of science curriculum reform in the United States. It is encouraging that political pressures from many sides have partially reversed that policy.

Fred Hechinger's article referred to above deplores the fact that opportunities for studying science in the schools are being drastically curtailed at present because of a shortage of supplies and equipment. These shortages were identified as the major problem in science instruction "by over one-third of the high school science teachers and more than half of the elementary school teachers in grades four to six. Without laboratory work, the N.S.F. report says, the enjoyment is taken out of science." What remains is a hard subject that students avoid, and guidance counselors steer children away from. The NSF-funded study referred to was carried out by the National Research Council by means of a task force. Its gloomy prognostications and its indictment of American school administrators who do not know how "to help teachers to develop effective science and mathematics programs" reinforces all I have said. The ominous consequences of governmental neglect cannot but reduce the United States to technological impotence in the face of more farsighted competitors.

Left to the fifty individual states, programs of reform in science education will not only be lacking in uniformity, they will probably, to judge by the past, not even exist. Perhaps only in individual universities can we hope to find some forward-looking development. Let us then dream a little and try to forecast what the university of tomorrow may become. Perhaps not more than half its total effort and operating budget will be spent on preservice education, whether for teachers, doctors, engineers, or other professional people. A gigantic new aspect of the university will be its devotion to periodic renewal of education, not in a haphazard way, but rather as a serious, and for many professions a mandatory, cyclical process. The costs of this new phase of education must in large part be borne by the employers, who will learn that profits accrue from continuously re-educated personnel. In respect to the schools, it will mean of course that support of such renewal programs must be met by the combined efforts of the university, the local school systems, the state, and the federal government.

To make of this enterprise itself a continuously developing, experimental part of the educational system, I can see three new types of organized effort, within a modernized School of Education. One of these units we may call an Instructional Resources Center. Here instructional design, use of visual methods and audio tapes, computer-assisted instruction, and other novel experimental educational methods may be combined with facilities for scientific testing of the reliability and validity of such methods. A second division we may call a Center for Curriculum Development, in which task forces of scholars and scientists together with experienced teachers give continuing thought to the design of new educational programs, especially those particularly needed for the renewal of education. In this Center the advanced methods of the new educational technology are to be brought together with the significant new developments of subject matter, and continuously new synthesis will be evolved for actual teaching. No such program would be worth much without the final practical test in the classroom and laboratory. The third organized unit, the Center for Continuing Education, is therefore necessary not alone to satisfy the needs for the renewal of education of professional groups and of adults in general but also to serve as the site where the new methods and ideas are put to the test. A steady flow of suggestions and criticisms from the staff and the students in the Center for Continuing Education will provide invaluable feedback for the

Center for Curriculum Development and the Instructional Resources Center. In terms of systems analysis, this should be a true cybernetic system. Here may well be born the fresh, vital influence that will ultimately sweep away the timeworn methods of instruction and the moldy subject matter that so sadly characterizes much of our university teaching.

It is not sufficient for the scientist in this day to train a new and more abundant generation to follow him. Surely, if the tempo of discovery and the alteration of ways of life continue, their incredible acceleration must place an utterly new demand upon education as a whole, and upon the scientist as teacher in particular. Can a civilization endure for long if it is based upon a scientific technology, while the populace grows ever more complacent in its ignorance of science, more superstitious about the methods which make science successful, more willing to leave direction and control to others? At the very least, no democracy can long survive on such a basis, for the power of government will either devolve upon unscrupulous demagogues or upon a scientific elite, so that tyranny or oligarchy must replace democracy, the self-government of an informed people. These questions of the future trouble many scientists and teachers today.

Of all the resources of a nation, its greatest are its boys and girls, its young men and women. Like other material resources, these can be squandered or dissipated. They are potential greatness, but only potentialities. Science creates knowledge and knowledge generates power; but knowledge resides only in the minds of persons who first must learn and be taught, and power is tyranny unless guided by insight and wisdom, justice and mercy. The greatest among us have been teachers, and the teacher is great among men.

9 The Human Multitude: How Many Are Enough?

Many scientists and teachers today believe that, next to the threat of the nuclear destruction of mankind in war, the curse of our enormously and seemingly uncontrollably expanding human populations throughout the world is the most critical danger to human well-being. Yet in considering the population problems of the earth at the present time, quality of population as well as quantity should be included.

QUESTIONS OF QUALITY AND QUANTITY

It is important to know not only what countries are growing most rapidly in population but also what differences exist in population density, in food and other material resources, and in the ratio of the producers to the two groups of non-producers—the young children and the retired adults. Demographic stress is related to the level of industrial developments achieved in a country, the kind of diet, the basic agricultural practices, the available capital for investment, and the recency of the abrupt decline in infant and childhood mortality. Population cannot be adequately considered without a study of the

---

This chapter was originally presented as an address to the annual meeting of the National Association of Biology Teachers in Denver on October 22, 1970. It was published, in an expanded form, in the *American Biology Teacher* in March and May of 1971 (vol. 33: 145–48; vol. 160: 265–69). The present revision contains added material, but on the whole what was said more than ten years ago is highly relevant and accurate today.

rural-urban shift (now being replaced in the United States by an urban-suburban shift). We cannot avoid the relation of population to pollution, to congestion, and perhaps above all to the psychological stresses growing from increasingly dense urban living. The level of education must be taken into account if an accurate prognosis of proposed measures to reduce population growth is to be achieved. Many factors enter into an answer to this question; is there an optimal size of population?

The quality of a population is not determined solely by its environment and its adjustments to that environment. The gene pool of the population establishes its ultimate potentialities. Although the gene pools of different countries are undeniably different in many ways, two general statements seem to be valid:

1. Inasmuch as every people has lived, until the present century, in a natural environment to which it has had considerable time (thousands of years, hundreds of generations) to attain evolutionary adaptation, it seems highly probable that each people, through the process of natural selection, is rather better adapted to its own environment than to any other that is quite different.
2. The adaptation of a population depends to a considerable degree on the diversity of its genotypes. Too great a uniformity might in fact prevent certain types of genetic adaptation that depend on the superiority of the heterozygous genotype or on the cooperative interaction of different genotypes.

The health achievements of the past century or two have largely eliminated natural selection based on differential mortality. We may add that differential fertility will also disappear as a significant agent of selection among genotypes whenever zero population growth is attained. Finally, we should emphasize that our present knowledge of the nature of the evolutionary process and our ability to control mutation, selection, and isolation of breeds, together with the possibilities of introducing eugenic measures of control and genetic manipulation, place in the hands of human society the power to control human evolution. In fact, quality control is hardly separable from quantity control, in the pursuit of population control in general.

Let us first consider the control of population quantity.

SOME FACTS AND EXPERT GUESSES

The facts and the projections are by now familiar to nearly every educated person. Starting from very low densities and very slow rates of increase in preagricultural times, world population reached about half a billion in 1650. It had doubled since Jesus' time. Then it doubled again in the next 200 years, doubled again in less than 100, and at the present rate of increase (2 percent a year) will continue to double every 35 years. As the world population has already passed 3 billion, the end of the century will likely see 6 or 7 billion persons in the world. The United States, at its modest present annual rate of increase (1.5 percent), will in less than a century have a larger population than the present population of China, now the most populous country in the world. China has a larger area than that of the United States; thus our population density will be greater than the average density in China today. Mexico, at its present annual rate of increase (3.2 percent), would possess, a century from now, over a billion persons. If these figures do not convince you that the present rates of increase cannot continue much longer, I might cite a calculation made by one of our leading demographers, Philip Hauser.[1] He pointed out in 1960 that even a 1 percent per annum rate of increase—a low rate, now attained only by Japan and the countries of Europe—is sufficient in a mere 5,000 years to create, from only 1,000 persons to start with, a theoretical population almost equal to the present world population of 3 billion persons on *every square foot* of the earth's land surface.

Even now, over half of the people in the world are badly nourished, and even more are chronically hungry. A doubling of the population in less than a quarter-century requires that the production of food be increased by at least that much or else the population will be even hungrier. Yet of the world's land area only one-fourth is arable. Much can be done, of course, to improve the state of food production and nutrition. Widespread protein deficiency may shortly be corrected by greater use of fish meal. By breeding more productive strains of wheat, corn,

[1]Philip Hauser, "Demographic dimensions of world politics," *Science* 131(1960):1641–47.

and especially rice—the staple food of over half the human race—and by effective fertilization, yields per acre (or hectare) can be increased sixfold or more. Yet I would caution those who, like some Russian scientists with whom I once discussed the problem, would claim that the only real problem is how to produce enough food. The United States is probably capable of producing enough food for 800 million people, if every possible improvement of production is utilized; but it cannot go on exporting large surpluses to more needy nations. And I doubt that Mexico, even though it has made such amazing strides in the past few decades in improving its production of wheat, beans, and corn, can adequately feed a billion people. Agriculture as we know it has certain upper limits of productivity determined by the land area. To be sure, long before a century has passed, and perhaps before the year 2000, the seas will be farmed more and more extensively, and a detailed knowledge of the chemistry of photosynthesis should make it possible for man to synthesize unlimited quantities of sugar from virtually inexhaustible carbon dioxide and water. Amino acids, fundamental to protein nutrition, are already being synthesized in limited amounts. Starches and other foods will come from sugar. The vitamins we can make already. Ultimately, then, it is not food so much as space itself that limits human population.

Megalopolis is an early, inescapable product of the population surge. Megalopolis grows not only because of the intrinsic increase of population, but also because of the increasing efficiency (including mechanization) of agriculture that is required if a larger population is to be adequately fed. Fewer people are needed on the land, and the cities teem with dispossessed millions who leave the countryside. Another of our principal demographers, Kingsley Davis, has estimated that in India, for example, this source of urbanization will bring 100 to 200 million people into the cities by the year 2000 when the largest city will contain 36 to 66 million people.[2] Clearly, the entire social order will have to change to provide a livelihood for these millions. The prospect is far more dismal than even the fate of those thousands dispossessed by political events in the Middle East.

[2]Kingsley Davis, "Population," *Scientific American* 209 (9:1963):62–71.

The first fruits of any marked improvement in the food supply of those countries now most rapidly increasing in population—in Africa, Asia (except Japan and the USSR), and Latin America—is almost certain to be a further increase in the birth rate. The population crisis will thus be aggravated, since it arises chiefly from the disparity between a rapidly falling death rate and, in most of the industrially undeveloped nations, a relatively stable birth rate. Yet these births, in view of the low infant and child mortality now characteristic everywhere, imply that in some twenty years there will be a more populous cohort of young men and women eager to marry and embark upon parenthood. A change in the death rate has an immediate effect upon the population, but if the change is largely an extension of the lives of older people there are few prolonged, secondary effects in later times. A change in the birth rate, however, has effects that spread out over succeeding generations in never-ending waves of influence. The baby boom in the United States in the 1940s produced massive, abrupt needs in the elementary and secondary schools in the 1950s and in the colleges in the 1960s, and even a secondary baby boom twenty to thirty years after the original.

Thomas Robert Malthus, in *An Essay on the Principle of Population* (1798), asserted that "population, when unchecked, increases in a geometric ratio" but "subsistence increases only in an arithmetical ratio."[3] That human populations are capable of doubling themselves in twenty-five years, as Malthus claimed, is amply evident at the present time. On the other hand, Malthus supposed that the greatest increase in the production of food could amount to no more than an amount *equal to present production* during each succeeding twenty-five years; hence the increase was arithmetical (as in the series 1, 2, 3, 4, 5...). The great modern development of scientific agriculture has proved Malthus wrong: at least for a time, the production of food can also increase exponentially. It is easy to see why this should be so, for food is composed of animal and plant products, and the

[3]Thomas Robert Malthus, *An Essay on the Principle of Population as It Affects the Future Improvement of Society*, 1798. Reprinted as "Population: The First Essay," (Ann Arbor: University of Michigan Press, 1959).

animal and plant populations, according to Malthus' own notion, ought to be capable of increasing geometrically (according to the series 1, 2, 4, 8, 16...). What really sets the limits of the growth of a population is the available space, or the accumulation of toxic wastes, or the exhaustion of fertility in the soil, or the waters in which grow the green plants, upon which all animal life depends. In the end, however, Malthus's conclusions are correct. Exponential growth in a limited environment cannot continue indefinitely. Sooner or later the increase is checked. In animals and plants this takes the form of the Darwinian struggle for existence and the survival of the fittest—though we now see that the so-called "fittest" are simply those hereditary varieties that under particular conditions are more viable or more fertile and outbreed their competitors.

SURVIVAL OF THE YOUNG

In human societies, for many millenia past, the principal factors that affected differential reproduction were those that modified the ability of the infant and child to survive the perils of disease and malnutrition and eventually to reach adulthood. Today the almost complete elimination of infant and childhood mortality in all countries that have introduced modern standards of medical care and nutrition has had striking effects. First, obviously, there has resulted the tremendous surge in population. Second, differential fertility has replaced differential childhood-survival as the chief basis of natural selection among hereditary human types. Both of these consequences involve psychological and social factors as much as biological.

Darwinian natural selection postulates the death of many young individuals, both before and after they reach maturity, mainly because of malnutrition, disease, and weakness in the face of predators. Yet even here psychic factors enter in. Any genetic predisposition that makes an animal sterile or that weakens its eagerness to mate will reduce its chances of producing offspring and consequently will reduce the relative frequency of its type in the population. Rats or mice confined in a pen with plenty of food and water and plenty of nesting material will multiply to a certain density, at which much fighting and terrorizing goes on.[4] The females then refuse to mate; or if mated, they frequently fail to become pregnant; or if delivering young, they fail to make proper nests, abandon their young, or even devour them. The population quickly fails to increase further and may even decline in numbers.

Man is possessed of sufficient knowledge and foresight to avoid such a predicament. He is nevertheless not exempt from the Malthusian dilemma, which foretold "vice" as well as "misery" as the chief causes curtailing population increase. By "vice" Malthus meant the consequences of purposeful human effort to restrain population growth. He had in mind the postponement of marriage to advanced years, as well as outright celibacy, with all the accompanying evils of prostitution and induced abortion. He might well have dwelt also on infanticide, for that crime on conscience has been a common social practice at various times. I recall my own boyhood years in China, when the raised stone platforms on which infants were exposed to the elements until death came to them were a frequent sight in the neighboring hills. The reasoning back of this seemingly heartless practice is truly illuminating.

The exposed infants were almost always females. The poor, in China, in the absence of any other form of social security, were utterly dependent on sons to support the infirmity of old age. It was common reasoning to say to oneself, "If I have no son to outlive me, what will I do when unable to work for my living? But to produce one son to outlive me, I must actually produce three or four, since few children survive the perils of infancy. Yet if I produce three or four sons, I am likely also to produce three or four daughters. These will not support me in my old age. They become the members of another family when they marry, although while unmarried they must be a charge upon my own family's scanty resources. It will be all my wife and I can do to feed four children and rear them to adulthood. If, then, the gods should will that girls be sent in place of boys,

[4] J.B. Calhoun, "Population density and social pathology," *Scientific American* 206 (2):139–48; R.L. Strecker and John T. Emlen, "Regulatory mechanisms in house mouse populations: The effect of limited food supply on a confined population," *Ecology* 34(1953):375–85.

we cannot rear them. We must therefore try to sell them to wealthy families as bondchildren; but alas, if we find no buyers—the market for girl children being so limited—we must leave them on the high places. The gods have given, the gods will take away. It is a sad thing, but we must have at least a surviving son!"

Many factors have contributed to China's enormous population growth in the present century. Not least among them was the abandonment of the practice of infanticide. The reduction in the number of females in the population had once been particularly effective in curbing the population increase. What has happened since infanticide was eliminated and the death rate of infants and children so greatly reduced? Without any change in the psychology of the masses, the ancient motivation to produce at least four sons has persisted. Now, however, they all survive to adulthood, and four girls on the average survive with them. It is clear as crystal that the psychology of the people must be fundamentally altered—a fact that explains why the Communist leaders of China are emphasizing in dire urgency the need, and admirable merit, of postponing marriage until the age of twenty-eight years or later, and why they are demanding the restriction of sexual activities and opportunities among students and other young people.

In India, too, it seems evident that no method of population control, so euphemistically called "family planning," can succeed without the introduction of a widespread, basic recognition of the existence of satisfactory social security to replace the peasant's need for sons to provide for him in his old age. However admirable and effective the new contraceptives may be, the steroid pills and intrauterine devices fail where they are most needed—not because of any lack of intelligence on the part of the peasants, but because the needs of old age can be satisfied only by sons. Even under present health conditions, one should be assured of at least two sons; the consequence is an average family size of four children. As our experience in this country shows, such productivity is more than sufficient to produce a 2 percent per annum rate of increase, doubling the population in thirty-five years. To me the conclusion seems inescapable: what India needs, even before it gets so-called family planning through contraceptives, is real family planning that provides a solid system of social security—a system in which men can put faith.

CONTRACEPTION AND CHANGING MORES

As human reproduction comes more and more under regular scientific control, we may expect present difficulties in the use of contraceptive methods to be greatly reduced. A recent promising advance, already successful in mice though not yet in humans, is that of placing a small capsule of progesterone under the skin of the female. Microdoses seep through the porous plastic into the circulation and prevent conception without side effects and even without modifying the normal female cycle. The capsule can be put in place easily and painlessly, and it may be removed at any time. This method may entirely supersede induced abortion and other less harmful but inconvenient methods of preventing implantation or growth of the fetus. Temporary ways of inducing sterility in the male without the accompaniment of side effects are also promising.

As countries grow more desperate in the face of a population increase that exceeds the growth of the food supply, we may be sure that old restraints will crumble, old freedoms and so-called "human rights" will become constrained. I can foresee a time when, in some countries, population has become so dense that no further increase can be tolerated. A marriage certificate might then bear two coupons entitling the couple to produce two children—no more. Taxation, graduated upward with increasing number of children, or sterilization, temporary or permanent, might be the penalty for exceeding the quota. Or—and here I speak hopefully as a human geneticist and eugenist—parents might be permitted to apply for an additional child if they can submit evidence that the two already produced are physically and mentally sound. That is a weak form of eugenic practice, but if carried out over a sufficient number of generations it might well improve the quality of the population.

Two other consequences of the inevitable population increase that faces most of the world in the near future must be mentioned. As the density of population increases and urbanization grows, to what extent will the population's mental health be impaired? Will we react to crowding more and more like the penned Baltimore rats and Wisconsin mice in the respective experiments of John Calhoun and John T. Emlen, referred to previously? I foresee a critical need for really effective regional planning, which would permit the enormously dense populations of tomorrow to have a frequent opportunity to escape the crowd and to commune with nature in field, forest, and wilderness. The present throngs of summer visitors to the Grand Teton and Yellowstone national parks, as well as most of the other parks, reinforce my conviction of the urgency of this national problem, even in today's sparsely populated United States.

Lastly, the conscious control of the increase in human population, by means of effective contraception, will challenge our sexual mores. Will promiscuity threaten marriage and weaken the stability of the family? I do not see how this can be prevented, and like Robert Morison,[5] I am fearful of the consequences unless some satisfactory social alternative to the family rearing of children can be devised. Throughout at least a million years, and probably far back in man's primate ancestry, the family group has been the basis of his personal security, the soil for his growing power, the source of his finest emotional qualities of love and altruism. The plight of the mother and the child lacking husband and father is exceeded only by the ultimate loneliness of the predatory male and the abandoned child. Humans will cease to be the *human* species when this has occurred.

IS THERE AN OPTIMAL SIZE OF THE HUMAN POPULATION?

This question poses to a geneticist the immediate need for definition of several terms.[6] First, of course, one must ask:

[5]Robert S. Morison, "Where is biology taking us?" *Science* 155(1967): 429–33.

[6]Bentley Glass, "Genetic and evolutionary considerations," in *Is There an Optimal Level of Population?*, S. Fred Singer, ed. (New York: McGraw-Hill, 1971), pp. 114–26.

optimal for what? Second, precisely what are the definitions of "population" and "size of population"? Third, is one to be concerned chiefly with the welfare of man in the immediate present or in the distant future? Finally, if the optimal size of the population is related to changing parameters of the environment, can we expect to preserve an environment sufficiently buffered against change that a prescription today will have any validity whatsoever for tomorrow?

CONDITIONS OF HUMAN ADAPTABILITY

A biologist must examine these questions in the light of "mankind evolving," to use Theodosius Dobzhansky's apt phrase. The primary characteristic of *Homo sapiens*, in contradistinction to other mammalian species and in particular his closer primate relatives, is the high order of intelligence he possesses—a quality that enables him to adapt his environment increasingly to his own desires instead of awaiting the far slower genetic adaptation of himself to his changing environment. Man's own adaptability, including his power to learn from experience and, through science and technology, to enlarge his power to exploit and modify the environment, depends on the considerable diversity of the human gene pool. It seems to me that although a "brave new world" is conceivable—that is, one in which man is artificially constrained to be genetically homogeneous or is reshaped into a few genetic castes—it is neither likely to develop within a democratic society nor is as optimal as a continuance of extensive genetic diversity. For mankind, a continuance of his adaptability and maintenance of the diversity of his gene pool would seem to be optimal, even to the remote future.

According to Sewall Wright,[7] the situation most favorable for progressive evolution is one that is provided by an oscillation between a relatively large population homogeneous in nature and a state of subdivision into many local, fairly isolated popu-

[7]Sewall Wright, "Physiological genetics, ecology of populations, and natural selection," in *Evolution After Darwin*, Sol Tax, ed., (Chicago: University of Chicago Press, 1960), pp. 429–75.

lations, at least some of which are of medium or small size. In that case the small isolates will come to differ from one another both in concealed genetic variability (genotypes) and in physical characteristics (phenotypes), and natural selection may discriminate between them. This view is generally accepted by population geneticists. Only in relatively small and isolated populations, because of inbreeding, do we find the genetic variability forced into expression and thus made subject to selection. On the other hand, if the total population of the species consists of only a few small isolates, there is too much danger that none of them can provide a phenotype permitting preservation in the struggle for existence. Here both chance and adverse selection will combine to magnify the likelihood of extinction.

In the one or two million years of existence of the human species, the total population was at first undoubtedly small, and it was also broken up into small, local breeding units that might be designated as bands, or extended families. Inbreeding was high; diversification among the isolates was probably great. The risk of extinction of the species was overwhelming until the total population of the species rose to 100,000 or so, when there would have been many bands. Later, as group methods of hunting developed and the usefulness of fire as a protective weapon was discovered, larger tribes succeeded the small bands. Ever since, we have moved steadily toward larger and larger Mendelian populations, with always a higher proportion of the variability in the gene pool concealed, and with less and less diversity of isolates and racial groupings. The human melting pot came into being. The melting pot has not yet eliminated all racial distinctions, but it is rapidly doing so. Genetic variability now chiefly differentiates individuals of the same breeding population, and a greater proportion of it is concealed in the recessive state. Populations have grown so vast that sheer distances, both geographic and social, have become primary factors in offsetting a purely random breeding pattern. Evolutionary change has thus slowed down, even without the effect of the diminished rigor of natural selection that has also taken place because of better standards of health and nutrition throughout the world. As the population structure of man has thus changed, his evolutionary prospects have altered.

FASTER CHANGE, SLOWER SELECTION

The present rate of change in the human condition is exceedingly great. No genetic change can occur fast enough to cope with it even under conditions of maximum mutation rates and rigorous selection. Mutation rates per gene are in general low, although the great size of present human populations ensures that many mutations, of each and every gene, will occur in every generation. The new genes will also undergo extensive recombinations, so that many different coadapted gene systems will exist. At the same time, our large populations ensure that most of the genetic variability will be concealed in the recessive state. Selection will consequently act mainly on the heterozygous effects of genes, which are commonly smaller in extent than the homozygous effects.

Selection has been further slowed down by the advances in medicine and nutrition that now virtually guarantee that every child born alive, and some that previously could not have been born alive, will reach the reproductive years. The reduced rigor of selection is increasing the abundance of once-detrimental genes in our population, for these genes no longer lower their possessor's chance of reproduction. Dominant genes that are reduced from $s = 0.5$ (elimination of half the genes in each generation) to $s = 0.01$ will tend to increase fifty times in frequency. Recessives undergoing a similar moderation of selection will increase ten times in frequency. Thus, in anomalous fashion, we have found a way to increase our genetic variability without requiring any alteration of the mutation rates—though these too may become greater through greater exposure in the modern world to ionizing radiations, chemical mutagens, and increased temperatures.

This genetic variability, however, is of little advantage to us in producing any real adaptation. We now modify the environment around us by technological means and create for ourselves a novel and artificial world in which our defective genes, even though active, fail to impede us in reproduction. The man of tomorrow will clearly need many more pills and prosthetic devices. That need not worry us so long as the social cost is not too great and so long as we can maintain our artificial environment unimpaired. By natural standards we may become degenerate physical beings, but we will not wish to apply natural standards any longer.

What is the optimal size of population? Man has achieved his present status chiefly through intelligence and adaptability. That evolutionary process occurred in populations very much smaller, and much more broken up into breeding isolates, than we now see. Evidently the great size of our present populations is not needed for progressive evolution and the maintenance of diversity. Our new condition provides us with added genetic diversity, but at a considerable cost: the social repair of genetic defects. This excess of diversity seems neither necessary nor advisable.

There is, I believe, only one condition under which it may actually save our species. If in the end we blunder into a nuclear war and thereby eliminate nine-tenths of the human lives on our planet, the scattered survivors will find themselves once more in the population structure of early Pleistocene man. Those who survive the direct effect of heat and blast and radiation and who are later spared by the fallout may still perish in large numbers because of their inability to provide themselves with the special foods, drugs, and devices on which they depend for life itself. Among the others, in remote inbreeding isolates, there will very likely be sufficient genetic diversity to enable *Homo sapiens* to make a wiser fresh start.

WHEN IS THERE OVERPOPULATION?

Indeed, when is there overpopulation? The question cannot be answered in simple terms. There is one answer to be given if you are thinking in the purely quantitative terms of numbers of people per square mile. But what sort of land is it? And what is the standard of living of the populations? Quite another answer must be given if you are concerned with the quality of the population in genetic and evolutionary terms, and not merely with the state of its affluence. I am personally ready to accept a thoughtful definition I received from John A. Moore in an unpublished manuscript:

> *Overpopulation* exists in a unit of society when that unit uses the earth's resources at a greater rate than they are restored by natural or artificial means.

In that case Americans already live in the most overpopulated land in the world, for they use and destroy, waste and dissipate the irreplaceable resources of the world at a rate that is estimated to be fifty times as great per person as it is for an Indian or African. Our enormous technological power is eating away the earth's limited supplies of fossil fuels and scarce metals at a rate that will exhaust them in a century, more or less. We devote little effort to the development of a new technology that will enable us to recycle and reuse materials to a maximum extent.

We are now entering a new era in the dimensions of human power—a day when we can modify the quality of the human population genetically as well as through improvement of the environment, a day when the course of human evolution will be placed in human hands. If we cannot solve even the relatively simple problems of controlling the quantity of human life on this so tiny planet, can we reasonably expect to manifest a greater wisdom in remodeling human nature and enlarging human capacities? We may foretell what man can be, but what *should* man be?

10 Developing Countries and the Explosion of Scientific Knowledge

The current exponential increase of scientific knowledge has profound effects upon science education in both highly developed and newly developing countries. It alters rapidly and profoundly both what knowledge is most applicable to human needs and what knowledge, from the standpoint of scientific understanding itself, it is most important to teach. It demands constant attention to the revision of the content of the science curriculum. It requires extensive alterations of the programs for training new science teachers. It makes necessary the continuous retraining of science teachers. It lengthens the period required for adequate training of research personnel and makes postdoctoral training a necessity. It thus contributes to the "brain drain," while at the same time the need for up-to-date, vigorous scientific leadership in each country is multiplied through the obsolescence of the existing cadre of university, high school, and elementary school science teachers.

The scientific needs of the developing countries must be assessed in the light of these facts. Among these needs the paramount ones appear to be as follows.

1. *To provide a good curriculum for training science teachers for the local educational system.* A well-trained science

---

This chapter was developed from two working papers presented for discussion at the sessions of the Fourteenth Pugwash Conference on Science and World Affairs held in Venice, Italy in 1965, and printed in the same year in the restricted Proceedings of the 14th Pugwash Conference, pp. 134–41 and 142–47.

teacher must know both *what* to teach and *how* to teach it. For the first, it is important that the curriculum should undergo continuous critical revision. Significant new knowledge, especially new fundamental concepts and approaches, must be quickly introduced, necessarily at the expense of less important older materials that must be pruned away. By many science educators it is now accepted as sound policy that the emphasis in teaching a science should fall more strongly upon the development of an understanding of the nature and processes of scientific inquiry and discovery, of methods of verification and experimental testing, rather than upon accumulated scientific knowledge as such. The science curriculum studies conducted during the past two decades in the United States, and now so widely adopted or adapted elsewhere, have in general pressed this point of view. For the general citizen, as distinguished from the practicing scientist, a knowledge of the place of science in man's culture and technology, and of its influence upon his philosophy of life, are even more important than a command of scientific facts and principles. It follows that, in instructing a young teacher in the matter of *how* to teach science, great attention must be given to the psychology of motivation and learning as well as to the specific methods of lecture, discussion, demonstration, and laboratory teaching.

In the work of the Biological Sciences Curriculum Study (BSCS), to cite one case, the leaders in the construction of the new curriculum have become convinced of the following points of view. (1) A good lecture is primarily of value for motivation and is of subsidiary value for conveying information. (2) Discussion is of value chiefly for clarification, but if "invitations to inquiry" are used to combine the analysis of a scientific topic with the study of a particular aspect of scientific method or philosophy, then discussion becomes far more valuable. (3) Demonstrations with models, actual materials, or experimental set-ups, either during lectures or with audiovisual apparatus at other times, satisfy the need to examine actual scientific phenomena and evidence but should occupy less of the laboratory teaching time and should instead be made more a part of assigned individual activity, comparable to reading a textbook. (4) The laboratory is best used for teaching if it is sharply focused on the processes that lead up to and are required for actual scientific investigation of a genuine problem with a real "unknown" parameter. (5) Students should be effectively trained to work in teams of two to four, each team doing a part of the entire experiment conducted by the class, which must thereafter have ample scope for presentation of results, discussion, comparison of controls with experimental systems, and synthesis of results. (6) Examinations can play an overwhelmingly destructive role in teaching by failure to direct the questions at the teaching objectives and instead emphasizing only factual information. Accurate scientific information is necessary but should be tested in the context of broader objectives, such as the ability to form a reasonable hypothesis on the basis of given data or the ability to derive a correct conclusion from the data. Finally, (7) experiments can be effectively performed by means of motion picture films that present data, allow formation of hypotheses, and permit further testing of hypotheses. Practice teaching, so widely and so properly required on the part of an apprentice teacher before he undertakes independent teaching, should be directed toward the use of these methods, as well as mastery of the new curricula in each respective field.

A good curriculum in a science must be continuously revised to keep up with the rate of advancement of that science. Thus, in very active areas of biology, a thorough revision at least every five years seems necessary. The doubling time of scientific knowledge in such areas is about ten years. Not only does the mass of information double, but the introduction of fundamentally new concepts and newly discovered laws, as well as the revision and correction of older, inaccurate beliefs must also be made at short intervals. It is important that students come to realize that a science is a growing, changing body of knowledge, not a set of authoritative dicta. In the experience of the BSCS, best results in revising the curriculum are obtained by teaming together scientists who know the advancing front of knowledge with skillful, experienced teachers from the schools who know the capacities and problems of the classroom.

A good curriculum is adapted to the local educational system. In biology this is necessitated by the differences in the faunas and floras, as well as the climatic environments, of different parts of the world. In the earth sciences, the local scene should obviously be the point of departure. In physics and chemistry, less local determination of content would appear to be necessary, yet even in these fields the choice among many possible examples and illustrations will be far more likely to arouse the interest of students and to involve them in active scientific inquiry if local situations and applications are utilized. For example, the production of nuclear energy in a developing country is less immediate than are the current resources of conventional fuels; and the teaching of physics, chemistry, and earth science should take account of immediate circumstances as well as long-range developments. Above all, the study of science in a developing country ought to emphasize the value of studying all parameters of complex systems. Many of our current ills, in highly developed countries as much as or even more than in underdeveloped lands, arise from the neglect of studying all parts of the system, a neglect resulting in a lack of foresight of certain adverse consequences. Nowhere is this more evident than in our failure to analyze adequately the effects of the output of human activities on the total environment. The useful products of industry may be outweighed in value by the destructive output of wastes freely liberated into air or water, or dumped on the land. The density of human population in an area not only establishes requirements for food, water, air, and goods, but also creates definite problems of waste disposal (sewage, garbage, air pollution, etc.). Long-range effects as well as immediate effects are rarely studied sufficiently. At all ages, school children and students must be taught to appreciate the existence of complex relationships in natural systems and the importance of evaluating human activities in terms of their total effect over time.

It is especially important that the curricula and the materials to be used in the schools and the spirit employed in teaching science may foster a more universal international understanding and friendship like that of the international character of science itself.

2. *To provide adequately for training sufficient numbers of science teachers.* In a complex modern society that is rapidly changing because of the technological developments based on scientific discovery, the cadre of scientists and engineers must be maintained at a sufficiently high level in numbers and training in order to promote further growth of the society. In

the United States, about 1 percent of the population is engaged in science and engineering. This figure includes the technical assistants, most of whom have a bachelor's degree in science, as well as the more advanced personnel. It is equally necessary to train the entire population in an understanding and appreciation of the major forces that today alter the conditions of man's existence and that change his view of his place in the universe. In the United States, where almost the entire population today completes a high school education, rather more than 1 percent of the population, or some 3 million students annually, should receive that sort of education in science. If now we assume that every elementary school teacher teaches elementary science (those who teach no science being approximately balanced by those who teach only or mainly science), we find a requirement for 960,000 elementary science teachers, or 0.48 percent of the population. In the secondary school (six years), science is taught mainly by specialists who teach only science, or 0.15 percent of the population. The actual number is probably considerably smaller, because I have assumed that students will take science during each year of the secondary school program, and many do not; and also because I have assumed that no science teacher can handle properly more than three classes a day of twenty-five students each, whereas many are required to teach four or five classes. The calculation is nevertheless useful in giving us a general idea of the proportion of the total population needed for science teaching in a population about one-fourth of whom are children in elementary or secondary schools. The total number of elementary and secondary school science teachers needed approaches two-thirds of 1 percent of the population.

If now we assume that the average length of service of one of these science teachers is fifteen years, the training program in the United States must annually turn out over 60,000 elementary school teachers trained in science as well as other subjects and 20,000 secondary school science teachers (junior high and senior high, grades 7–12). In the United States, this is about one-fifth of all the new secondary school teachers graduated from college each year. If the average length of service is less than fifteen years, the number trained annually must be greater. Statistics on length of service in the profession are not readily available. Although teaching standards in the United States are by no means optimal, the magnitude of the training program required to attain a similar educational standard may be simply calculated from the proportion of each country's population to the 210–220 million persons of the United States.

3. *To provide science education centers to retrain science teachers continuously as a regular part of the teacher's annual duty.* In 1966 I described the Japanese Science Education Centers. At that time I had an opportunity to visit a number of these centers and to observe them in actual operation, some six years after the first ones were established. They have been so remarkably successful in providing retraining for science teachers on a local basis, as a part of the prefectural school systems, that a large proportion of the Japanese science teachers, both elementary and secondary, have in the past two decades been materially aided by them. Their success in the improvement of instruction has been so evident that the Ministry of Education in 1965 embarked upon an extension of the plan to include the other subjects commonly taught in the schools.

The Japanese plan has several admitted weaknesses: first, university faculties are very little involved in the programs of the centers; second, the facilities do not permit sufficiently frequent and prolonged study at the centers by each teacher; and third, there is inadequate staffing to permit effective planning of the courses given for the teachers, who definitely are neither undergraduate students nor typical graduate students specializing for a doctoral degree. The Japanese have obviated the last-mentioned of these deficiencies by making critical use of the various curriculum studies developed in the United States by teams of scientists and teachers supported by funds granted by the National Science Foundation. And they have improved upon them. Some universities in the United States, Great Britain, and no doubt elsewhere are endeavoring to remedy the first of the three deficiencies by making the university itself the location and planning center of the retraining program. To overcome the second of the three deficiencies will unquestionably require the most determined and persistent large-scale effort. The cooperation of the school administrations in releasing teachers from classroom duties for retraining programs is essential to it; the financing of large-scale programs by government aid is equally necessary.

Joint planning and international cooperation in these programs remains the best way to cope with the obsolescence of science education. I wrote in 1965, "Permanent, continuous study groups should be formed to correlate the ideas now being generated in different countries in regard to science education, and to inform and advise all countries of the best new plans and programs."[1] UNESCO should be stirred to more vigorous action in these matters, along the lines of its praiseworthy support of special training programs in Asia, Africa, and Latin America, but less through sporadic conferences and local curriculum projects than by establishing and supporting international study groups and clearinghouses, so that all successful developments anywhere may become rapidly utilizable for adaptation everywhere.

Science education is in a desperate state of obsolescence in even the most favored countries of the world. No teacher can afford to go for more than five to eight years without extended retraining. Most countries have made only halfhearted attempts to do anything at all to meet this need. How strange that although the economic theory and practice of all countries accept the idea that man-made structures require maintenance and that charges for this are regular features of budgets, no nation seems fully to have accepted the idea that the training of its public servants undergoes an obsolescence proportional to the rate of change in society, to the development of technology, to the advances of science! The rotten house will fall. It is no less certain that the decay of knowledge and education will rend the fabric of society, whether socialist or capitalist.

4. *To send a sufficient number of the best students abroad to engage in advanced scientific study.* The developing countries need certain numbers of highly trained scientists and engineers for the growth of their own institutions and industries. Insofar as these cannot be trained in the home country, they should be sent abroad to the best schools. In the past, support for fellowships abroad has often come from foundations and foreign governments eager to receive and train the able leaders of countries whose development they wish to assist. With current

[1]Bentley Glass, "Editorial: Japan points a way," *Science* 150(1965):1107. Bentley Glass, "The Japanese Science Education Centers," *Science* 154(1966): 221–28.

rapidly growing demands on all such funds, it is evident that either the developing countries will themselves have to assume a larger share in what indeed is a future investment in their own development, or the supply of scientists and engineers they obtain will fall far short of their needs.

A highly developed industrial country profits by having about 1 percent of its population in scientific research, science education, and technical support. It may therefore be suggested as a goal that a developing country might send at least one-half of 1 percent of its promising students abroad to complete graduate education in universities and technical institutes. If these young people return to their native country at the expiration of their training abroad, experience shows that the national investment in their knowledge and skills will be repaid manyfold.

5. *To supply posts and research facilities in sufficient number and of sufficient attraction to bring most of these scientists home after training.* Unfortunately, most of the developing countries at the present time are seemingly unaware of the need to provide sufficient suitable openings for their best young people trained abroad. The result is the well-known "brain-drain." It is necessary to plan for an expansion of local scientific laboratories and university and college teaching posts in order to accommodate these future leaders of the nation in science and education, industry and government. Few of them would elect to stay in the countries of their specialized training if the opportunities back home were adequate. Many, in fact, return home and shortly thereafter, from lack of positions or facilities, are driven away again. A post with the equivalent of a few thousand dollars for research may be sufficient to hold a fine young scientist. A junior post in a college or university will satisfy the educator if he is convinced there is no room at the top. Too many countries blindly follow a policy, imported from nineteenth century Europe, of professorships limited in number to one in each department or discipline, so that the young person knows at once that there will be no advancement in rank or income for many years.

My personal experience as president of a small American foundation that has made grants to young scientists abroad provides ample evidence that small grants supplied directly to young scientists and teachers often serve to make it possible to buy equipment, start research, support a student or hire an assistant, and begin productive scientific work when the university administration is powerless or unwilling to assist. Money poured into the bureaucracy from the top is usually dissipated, often without trace. Infused directly where needed, it is like water at the roots of a wilting plant. Is there any sound reason why governments themselves cannot plan and accomplish the same results? Why plant the young tree if it is not to be watered? Why not fertilize it when it is recently transplanted? Why not plant it where there is room to grow? Are we to be more sensible about trees and crops than about the infinitely more productive cadre of young scientists and science teachers?

6. *To establish an agency to make a continuous study of the relevance of new scientific discoveries to social, economic, and health needs of their own country and culture.* This final provision is doubtless the most important of all, for unless it is assured all else will ultimately fail. The curriculum for the training of science teachers, the adequate volume of new science teachers, the foreign training of the future scientists and engineers and teachers of teachers in the country, and the provision of suitable posts and resources for the scientific leadership when they return from training abroad will all be of little avail if that nation makes no continuing study of the relevance of the scientific and technological changes in the world to its own needs and the impact upon its own culture. The task, in other words, is to coordinate all of the preceding measures with the changing order of the world and especially with the economic pressures and social ferment of the home country.

Man is a social animal in a complex environment. At his peril he ignores the balance between density of population and its mode of life, on the one hand, and the resources of the environment, inanimate and living, on the other. Exhaustion of the resources may be less to be feared than the pollution of the air, water, and soil with inorganic and organic wastes. Nature's cycles are mighty but often delicate and subject to collapse when overstressed. Civilizations before now have destroyed themselves through waste of their soil or through passive encouragement of seedbeds of disease. It can indeed happen again, through ignorance or folly. The developing nation that is wise will therefore study closely all trends in the highly industrialized leading nations of our century and avoid their mistakes. Its scientific planners will take precautions, harbor resources wisely, prevent exploitation, and develop an educational system appropriate for both immediate and future needs. The growth of the population must be adjusted to the increase in gross national product and must take account of the ultimate limitations of space within the nation's boundaries. The ideal is a people adjusted to its land and its resources, daily increasing its richness of life through the applications of science and technology to its needs.

Those needs are unique and grow out of a nation's past as well as its present. They are shaped by its culture and its ideals as well as its material resources. As in the evolution of an animal species, its various aspects must change in consonance with each other, for too great a change in one respect, with too little in another, may lead to extinction. The science of tomorrow is truly to be the science of systems analysis, at present quite rudimentary. The proposed national agency is needed to relate the factors of change inherent in scientific discovery and education to the whole system of a people living in its homeland, and hence to plan intelligently for its future well-being.

11 The Biology of Nuclear War

Perhaps the most significant aspect of this essay is that it was first prepared more than twenty years ago, yet needs only minor revision to be valid at the present time. To a very great extent that fact is attributable to the lack of technological progress in the essential nature of the nuclear fission and hydrogen bombs. The data on fire and blast effects, and of course upon the nature and consequences of fallout, all date from the period of atmospheric weapons testing, which has been in abeyance since 1963 except for minor, rather infrequent tests by France in the Pacific, China in its far west, and possibly a few other newcomers to the "Nuclear Weapons Club." In the main, the result has been a great improvement in the size and accuracy of delivery vehicles, and of what George F. Kennan has recently called, very aptly, the "preposterous accumulations of overkill."[1]

On the other hand, there has been a steady and considerable increase in our knowledge of the consequences of low doses of ionizing radiation and the probabilities of the induction of mutations and cancers by such doses in human beings. That new information serves to strengthen the conviction that all high-energy radiation produces harm, in amounts generally in proportion to the accumulated dose. There is no harmless dose of high-energy radiation. Even the smallest dose increases

---

This essay was formulated as an address to the National Association of Biology Teachers at its annual meeting in Denver in 1961 and was printed in *The American Biology Teacher* 24 (1962): 407–25. It has been revised and is here reprinted with permission. Special thanks are due to Dr. Mary Esther Gaulden and her colleagues of the University of Texas Health Center in Dallas for their invaluable suggestions made during the revision.

[1]George F. Kennan, *New York Review of Books*, January 21, 1982, p. 10.

somewhat the probability of cancer in later years, after a latent period of two decades or more in most cases. That truth does not deny that, whenever the doses are small, the dangers are less than those of many everyday risks we run.

The sheer numbers of stored nuclear weapons represent the most incredible development of the arms race in human history. When 10,000 megatons, accurately targeted, is enough to destroy the military capability, and indeed the very life, of any major power, why are twice that or ten times that needed? The overkill is more and more preposterous when one contrasts the older times of bomber-delivered bombs to the present era of multiple-headed independently targeted reentry vehicles (MIRVs) of intercontinental range and of accuracy at 5,000 miles of a hundred yards or less. Yet every expert admits that interception and destruction en route of all the MIRVs that might be launched is quite impossible. The total array of intercontinental ballistic missiles, bombers, and special submarines like our Tridents guarantee the delivery on target of a substantial portion of the bombs launched. This is especially true because of the agreement to ban research and development in the area of antimissile weaponry.

We may therefore proceed to restudy the old analysis of the question with only minor new information to employ, and only minor alterations in the conclusions are to be expected.[2]

HOW BIG A WAR?

All sorts of magnitude have been envisaged for nuclear wars, ranging from minor, limited uses of such tactical weapons as a single man may carry and fire from a bazooka, to the worldwide devastation and fallout that might result if enemy powers were to launch 30,000 megatons of nuclear weapons, or more, at each other. The first question that must be answered in a discussion of the effects of nuclear war upon living things is simply: "How big a war?"

[2]The very recent, popular book by Jonathan Schell, *The Fate of the Earth* (New York: Knopf, 1982), reiterates without contradiction many of the facts and views set forth here.

A second question is of almost equal importance, though it presupposes that the nuclear war will not be of a magnitude that is all-destroying. This question involves the strategy employed by the military forces in their selection of targets. Will the nuclear weapons be aimed solely at the striking force, the "counterforce" of the enemy, that is, at its missile launching sites, its intercontinental bomber bases, and its other retaliatory forces? Or, on the contrary, will the weapons be aimed at the cities and industrial plants, the communication centers and the hubs of transportation of the opponent? The choice of one of these alternatives matters immensely, since the patterns of destruction from blast, fire, and high-energy radiation will differ exceedingly depending upon the location of the targets within a country.

It is consequently necessary for us to delimit our consideration of the problem by making tentative answers to these questions, even though we must keep in mind that the problems we explore and the conclusions we reach may be very different if different assumptions about the magnitude and the strategy of nuclear warfare be chosen. In the hearings on the "Biological and Environmental Effects of Nuclear War" held by the Joint Committee on Atomic Energy of the United States Congress in 1959, an attack on the United States of 263 weapons delivered on 224 targets and amounting to 1,446 megatons of explosive power distributed in weapons of 1, 2, 3, 8, and 10 megatons (MT) each was assumed as a basis for the calculation of effects. (The reader should note that a megaton, abbreviated hereafter as MT, is *not* a unit of weight. The term refers to an explosive force equivalent to 1 million tons of TNT. In comparison, the relatively tiny weapons exploded at Hiroshima and Nagasaki were reckoned at 15 to 20 KT, or thousands of tons equivalence to the explosive force of TNT.)

No specific reasons for assuming this magnitude of attack were given, but the description of the exercise indicates that an attack of this size would be sufficient to blanket the principal military counterforce targets in the United States. Included were 111 air force installations, 71 critical target areas (centers of population and industry), 21 atomic energy installations, 12 army installations, 5 navy installations, and 4 marine corps installations. The assumed attack was thus not directed *primarily* at population centers. For purposes of calculating the worldwide fallout, it was assumed that the attack on the United States was matched by the detonation of 2,500 MT elsewhere, so that the total fallout would result from the explosion of approximately 4,000 MT of nuclear weapons. It was further assumed that 50 percent of the explosive power would be derived from fission and the other half from fusion and that the radiactive products would be derived entirely from the fission reactions. This was at a time before it was recognized that the carbon-14 produced by the fusion process would constitute a considerable and very long-lasting portion of the total fallout.

No subsequent studies seem to differ except in scale. It is therefore convenient and valid to use this early study as a basis of analysis and simply multiply by whatever scale of exchange is now thought to be realistic.

Already in 1959 some of the effects of much larger attacks than the one assumed for the hearings had been calculated in various studies. One paper, reprinted in full in the published volume on the hearings, in fact calculated the casualties and deaths to be expected in "unprepared" contrasted with "prepared," that is, adequately sheltered, populations for attacks up to a level of 50,000 MT.[3] It seems probable that the much smaller magnitude of the postulated attack was dictated partly by the minimal size of attack needed to destroy the prime targets and partly by the desire not to alarm the general public by imagining an attack of far greater dimensions. It did not reckon with the need to allow for misses, and it was apparently not related to the existing stockpiles of nuclear weapons and the delivery capacities of the opponents, which were far greater than were needed to mount 1,500 MT attacks. These points seem to have been realized in planning the hearings held in 1961 before the House Committee on Government Operations,

[3]Hugh Everett, III and G.E. Pugh, "The distribution and effects of fallout in large nuclear-weapon campaigns," *Operations Research*, 7(1959), reprinted in *Biological and Environmental Effects of Nuclear War*. Hearings before the Special Subcommittee on Radiation of the Joint Committee on Atomic Energy, Congress of the United States, 86th Cong., First Session on Biological and Environmental Effects of Nuclear War, June 22–26, 1959, (Washington, D.C.: United States Govt. Printing Office, 1959), pp. 859–81.

later occasion a wide range of attacks, from 3,000 up to 30,000 MT, was postulated.

The casualties and destruction that would result from even a "small" 1,500 MT attack on an unsheltered population are appalling. The 1959 hearings had predicted that deaths would amount to 42 million persons in the 71 critical target areas, or 27.8 percent of the 1950 population on which the estimates were based. In addition, 17.2 million survivors would be injured. The same "modest" attack would produce today—under the same conditions, since no real increase in effective shelters has been produced—over 63 million dead and 25 million injured. Yet we cannot assume that even these figures represent a reasonable estimate of the disaster.

A simple multiplication in scale to an attack of 25,000 MT on the United States leads to the conclusion that the entire population would be wiped out, many times over. Only personnel sheltered adequately in deep rock, blastproof shelters could reasonably be expected to survive.

One strategy of attack assumes that the blow will be strictly against the retaliatory power of the victim. Another, however, assumes that the primary target will include the general population and the industrial centers of the enemy. It is hard to suppose that once combat has been initiated the disorder and failure of communications, together with the uncontrollable emotions which can be foreseen, will not lead to an all-out attack. Nor is it reasonable to suppose that, in the anticipation of a nuclear attack upon the Strategic Air Command (SAC) bases of this country, the military commanders would not deploy forces and thus make of every commercial airport for jet planes a logical target. These airports, of course, are in close proximity to all our major cities. Cities are also the communication and transportation hubs of a nation; and if the enemy would have reason to suppose that a first strike might not annihilate the counterforce and that a second counterstrike might be expected, then every center of communications would become a logical target. Long ago General Lauris Norstad, when commander in chief of NATO, said before Congress:

> I do not agree with those people who say you can control the size of this fire, the size of this blast, neatly, cold-

> bloodedly, once it starts. I think it is the most dangerous and disastrous thing in the world. I think you must prevent the thing from starting in the first place, because once it starts in a critical area such as the NATO area, it is more likely than not, in my opinion, to explode into the whole thing, whether we like it or the Russians like it or anybody likes it.[4]

It seems more reasonable, then, to consider the possibility of a nuclear war that would strike both at the counterforce and at the population, one that would attempt to deal so crushing a first blow that no second strike in return would be possible.

During the twenty-two years since this analysis was first written, the emphasis in the planning of nuclear war and defense has inexorably moved from considerations of the numbers and sizes of the nuclear weapons themselves to the development and deployment of the vehicles for their delivery. Bombers have been supplemented and now almost superseded by the invention and improvement of the intercontinental ballistic missiles, the equipment of each with multiple independently targeted reentry vehicles, the hardening of the silos in which the missiles are deployed by steel and concrete, and the rapid improvement in the accuracy with which the missiles can be directed to their targets. Nevertheless, the accumulation by both superpowers of nuclear weapons, both deployed and in reserve, has hardly diminished. A common estimate is that 15,000 to 25,000 megatons of nuclear weapons are now in existence.

We arrive at the conclusion that a 10,000 MT attack on either side, or a total 20,000 MT exchange, is not unlikely if nuclear war does occur; and the United States is far more likely to be hit with 10,000 MT than with 1,500 MT of nuclear bombs. An estimated 300 MT assigned for the destruction of each of about 20 missile launching sites, and considerably less per SAC base, might require perhaps 7,000 MT for the strike against the launching sites of our counterforce. The remainder, some 3,000 MT, is between five and six times the level of attack assumed in the 1959 hearings to be directed against the 71 critical target

[4]Quoted from *Fallout*, John M. Fowler, ed. (New York: Basic Books, 1960), p. 184.

areas, representing the major concentrations of population and industry in the country and ranging in size from New York City to York, Pennsylvania.

From the calculations previously referred to, we can now calculate the probable deaths according to various geographic distributions of the detonations and upon the assumption that fission and fusion are equal. Everett and Pugh assumed that the fission yield was two-thirds of the total yield; but bombs are "cleaner" now. A relatively unprepared population would then stand to lose about 82 percent of its numbers by death within 60 days and a total of 90 percent casualties if the density of the yield was proportional to the distribution of the population, or 60 percent deaths within 60 days and a total of 78 percent casualties if the yield density was proportional to the distribution of military air bases. These figures provide us with limits. Perhaps an intermediate figure is more realistic: about 150 million dead within 60 days and 35 million additional casualties. That is to say, of the 70 million survivors about half would be seriously injured. Furthermore, as Everett and Pugh stated, the casualties at 60 days are not the full toll of ultimate casualties. "Such delayed effects as the disorganization of society, disruption of communications, extinction of livestock, genetic damage, and the slow development of radiation poisoning from the ingestion of radioactive materials may significantly increase the ultimate toll."[5]

In the USSR, because the area of the country is so much greater and so much of it is sparsely inhabited, the effects of a similar attack would be somewhat lessened. Nevertheless, 62 to 70 percent of the population would die within 60 days after a 10,000 MT attack.

BIOLOGICAL EFFECTS OF NUCLEAR EXPLOSIONS

The destructive effects of nuclear detonations result from blast, thermal radiation (heat), and high-energy radiation (comparable to X-rays). Radioactive particles are also produced, carried up by the fireball, and distributed to a distance. These,

[5]Everett and Pugh, op. cit.

when they descend, constitute the fallout. It is probably surprising to most persons to learn that no one was killed at Hiroshima or Nagasaki from fallout. Those bombs, so puny by modern standards, burst in the air, and the one-mile radius of their blast and incendiary effects extended just as far from the hypocenter (directly under the point of detonation) as did their radiation effects. Air bursts produce relatively little radioactive material in comparison with surface bursts, and a 20 KT bomb does not release sufficient energy for the fireball to rise into the stratosphere and achieve worldwide distribution of fallout. Nor was there any significant local fallout, beyond the area of blast and heat damage. Fallout, in fact, came into the world as a virtually new phenomenon when weapons tests of megaton bombs began in 1952.

The biological effects of nuclear war would involve the several effects of blast, heat, direct radiation, and fallout. From bombs of moderate size, up to 1 MT, damage to living things is equally from blast and from heat. Above that size, the radius of damage from firestorm increases more rapidly than that from blast. For a 20 KT bomb, the radius of firestorm as well as that of blast is about one mile, as already mentioned; but for a 20 MT bomb, the blast radius of which is ten miles, the incendiary radius is thirty miles; for a 50 MT bomb, the blast radius is thirteen miles, but the incendiary radius is fifty miles; and for the latest giant, the 100 MT bomb, the blast radius is seventeen miles, but the incendiary radius is seventy miles. The blast radius increases as the cube root of the energy in megatons, but the incendiary radius increases as the square root. Gerard Piel put it tersely: "The blast effect would destroy the central city, but the firestorm would incinerate the metropolitan area."[6]

Consequently, for large air-burst bombs, the principal danger is that of fire and firestorm. The latter occurs when raging conflagrations over great areas so deplete the air of its oxygen that much carbon monoxide is produced. In World War II, in the great firestorms produced by old-fashioned incendiary bombs in Dresden, Hamburg, and Tokyo, far greater

[6]Gerard Piel, The Illusion of Civil Defense (New York: Scientific American, 1961), 23 pp. Reprinted in *The Fallen Sky: Medical Consequences of Thermonuclear War*, Saul Aronow, Frank R. Erwin, and Victor Sidel, eds. (New York: Hill & Wang, 1963), pp. 57–73.

loss of life occurred than in Hiroshima or Nagasaki. The deaths were estimated at 300,000 in Dresden, 70,000 in Hamburg, and 200,000 in Tokyo. Here, the occupants of bomb shelters were found suffocated and cremated in their shelters of refuge. The plain lesson is that shelters, to be effective in a metropolitan area, must be more than blastproof. They must be completely sealable and heatproof and must be provided with means of supplying adequate oxygen and removal of carbon dioxide for periods of twenty-four hours or more. Fallout is simply not the problem in the primary target areas.

Large bombs burst on the ground or in the water are another matter. From ground-burst bombs the main danger is from the firestorm, up to the limits of the incendiary radius. Beyond that distance, the danger is from the local fallout, composed of the heavier particles of radioactive material that are produced by the bomb burst and that settle out relatively quickly. These are particles from the size of cinders down to the size of ordinary house dust. Crooked air intakes for the shelters will keep out the bigger ones, and furnace or air-conditioner filters will eliminate most dust-sized particles, as most of us know from everyday experience.

THE EFFECTS OF LOCAL FALLOUT

The fallout contains radioactive isotopes of many sorts, some with relatively short half-lives, some with very long ones. Thus iodine-131, which has a half-life of eight days, is reduced in one month to less than 10 percent of its original amount, and in eight months it is reduced to less than 1 percent of the original amount. Though shortlived, iodine-131 is particularly dangerous because it is concentrated by the human body in the thyroid gland, where its radiation may kill some or all of the cells of the gland and produce serious or fatal consequences. Strontium-90 and cesium-137 last longer, their half-lives being respectively twenty-eight and thirty years. Strontium-90, resembling calcium in its chemical properties, is stored like calcium in the bones of the body. Here it emits beta rays of low penetration (about one centimeter), so that strontium-90 constitutes a hazard primarily to the bone cells and the bone marrow. Bone cancer and possibly leukemia might result, though neither would arise until long after the passage of the sixty days used earlier in reckoning the deaths and casualties attributable to nuclear war. Cesium-137 seeks out no particular tissue, but it emits highly penetrating gamma rays as well as beta rays. It thus constitutes more of a hazard to the entire body, including the reproductive organs where the cells reside that harbor the hereditary materials which will endow future generations. Fortunately, whereas strontium-90 is likely to stay in the skeleton for years or even a lifetime, cesium-137 moves rather quickly through the body, with an average biological residence of only seventeen days.

Carbon-14 must also be mentioned. Unlike the other radioactive isotopes already described, this one occurs naturally, being formed in the upper atmosphere by cosmic ray bombardment. Also unlike the others mentioned, it is formed by nuclear fusion rather than fission. What has been formed by nuclear explosions to date about equals the amount normally present in the atmosphere and is far less (about one-eightieth) than the amount present in the ocean, in plants, and in the air. We might therefore neglect this small increment as being of little importance, were it not for two things. First, carbon-14 has a very long half-life of 5,600 years, so that whatever is produced by man's nuclear explosions will remain over thousands of years. Secondly, though the biological residence of carbon-14 is only six months on the average, it can become incorporated into the very molecules of deoxyribose nucleic acid (DNA) which form the genes. When such an atom of carbon-14 disintegrates, it produces the destruction or gross alteration of the molecule of which it was a part. In hereditary terms, this means that the gene is lost or mutates, generally to a form that produces damage in some descendant.

The local fallout, because of irregularities of the wind, usually assumes the form of a twisted plume extending downwind from the point of detonation. In the case of a large bomb, it may extend 150 miles downwind and 50 miles across. The fallout is of course heavier in the center of this plume than toward the edges or the tip. In the case of a 10,000 MT war, these plumes of fallout would virtually blanket the country, especially in the area lying east of the Mississippi River. The intensity of radiation from the fallout drops off with surprising rapidity, roughly tenfold for every sevenfold increase in time. After six months, it drops off even faster. Thus, as calculated by Ralph Lapp,[7] a dose rate of 4,000 Roentgens(R)* per hour for the first hour following a detonation will drop to 2,500 R/hr. during the second hour and 550 R/hr. during the fifth hour. The total for the first day would amount to 12,150 R, but for the second day it would drop to 950 R. By the fourth day the total dose for the day would no longer exceed the median lethal dose for human beings. On the seventh day, with a mean dose rate of only 5 R/hr., the total dose for the day would be 120 R. At this level it would not be fatal for a person to stay outdoors and unsheltered for a considerable time, provided the total dose accumulated did not exceed 120 R. Even an initial dose rate many times higher than that assumed in these calculations would require only a very few additional hours to drop to the first level postulated here. Moreover, higher initial dose rates may be largely ignored, since they will occur only inside the radius of total destruction by blast and fire.

For most biological effects, provided the dose exceeds a minimum amount, the consequences are proportional to the total accumulated dose of high-energy radiation. That is, such biological effects as the killing of cells (which, depending on the nature and number of the cells killed, can have drastic effects and even kill a person) are proportional to the total accumulated dose above a certain minimum, or threshold. For genetic effects, however, produced in the reproductive cells and passed on through them to new generations, there seems to be no threshold of effectiveness whatsoever. That is to say, even the smallest increments of dose play a proportional part, so that

[7]Ralph E. Lapp, "Nuclear war," in *Fallout*, pp. 153–73.

* A Roentgen (R) is the physical unit of ionizing radiation. The unit of absorbed energy is the *rad*, which varies with different types of ionizing radiation, such as beta rays, gamma rays, or high-energy particles. It also varies for different plant or animal species, and therefore, whenever the discussion bears upon the effects on human beings, the technical unit is the *rem*. For general purposes, these three units may be taken as equivalent; or, whenever there is a material difference between them, that should be specified. For purposes of comparison, one may note that an average chest X-ray exposure delivers 20 to 50 millirads (or thousandths of a rad) to the skin, while the deeper tissues of the body receive less and less, the deeper they are, because of the shielding and absorption by the tissues above them.

the total accumulated dose is the measure that must be watched. In areas of heavy fallout, shelters should be constructed to provide a hundredfold reduction in the fallout intensity. This should generally suffice to keep the first-day dose below 100 R. The basement shelter that provides a protection factor of 10 or 20 times the safety of outdoor exposure is not adequate for this, although such shelters would of course be sufficient in areas where the fallout during the first day was much less, not exceeding 1,000 to 2,000 roentgens (or rads). If the exposure during the first day can be kept under 100 rads, for the first week under 150 rads, and for the first months under 200 rads, acute radiation sickness could probably be prevented, since much of the dose would be accumulated at a relatively low dose rate, which is less effective in producing biological damage than a comparable dose delivered very quickly. For the purposes of the following discussion, however, we must suppose that many, if not a majority of, survivors after a 10,000 MT attack on the United States would receive a total accumulated dose exceeding 200 rads. During the second month the outdoor dose might still exceed 200 rads and during the third month 100 rads, so that decontamination procedures and indoor living for much of the time would still be required.

In the 1959 hearings it was stated that the postulated 1,500 MT attack would destroy approximately one-quarter of all the dwellings in the country and would damage another 18 percent of them sufficiently to require evacuation for major repairs. Moreover, about 5 percent of the undamaged dwellings would have to be evacuated for several months to a year because of heavy fallout contamination. Without making detailed calculations it is evident that a 10,000 MT attack would destroy or severely damage most of the dwellings except in areas remote from cities and military targets. Those who survive in shelters must therefore expect to live in them for prolonged periods beyond the two to three weeks of acute fallout danger.

ADEQUATE SHELTERS?

From these considerations we can come quickly to some idea of what sort of shelters would really be effective in the event of any such war. An adequate shelter outside the metropolitan area should be fireproof and should be provided with proper ventilation and air filtration for the expected number of inmates. It should have its own private water supply, such as a well, and its own sanitary and bathing facilities; food and medicines for a stay of two weeks per person; battery-operated radio and light system; devices to monitor radiation, decontamination equipment; fuel for cooking and heating; and sufficient clothing per person. Not least important are the psychological preventatives of boredom and claustrophobia—games, books, magazines—for if the shelters ever need to be used there will be no television programs to watch and no radio programs to listen to except military and civil defense bulletins. What most people fail to realize is that there will be no electric current, that conventional sources of light and heat will be unavailable.

Inside the incendiary radius extending outward from each detonation but outside the blast radius, shelters must have additional features if they are to serve to preserve life. Protection against fallout alone is inadequate here. Even beyond the blast radius the shelters must withstand pressures up to five pounds per square inch. They must be insulated and be tightly sealable and have their own oxygen supply and cooling system sufficient for the duration of firestorms. Food and medicines should be provided for at least one month's needs per person. Perhaps the most important requirement here is that of trained personnel able to monitor the levels of radiation inside and outside the shelter; able to run the machinery needed to supply such a shelter with ventilation and air filtration, oxygen supply and carbon dioxide removal, lighting, heating; and able to give medical services. Under such conditions community shelters and a high degree of organization and training seem indispensable.

The provision of such shelters is not impossible for a nation. Within limits, Sweden has done so already for a portion of its population. To do so is, however, enormously more expensive than any program currently contemplated by our own government. Even a shelter outside the metropolitan areas and such as the one described, suitable for four to six persons, will cost upwards of $100,000. This statement is made on the basis of the experience of a fallout expert who actually constructed such a shelter in the Maryland countryside. Yet, it must be emphasized, no simple construction of railroad ties and sandbags can be of much value unless built in a very remote place and unless the powers agree to engage in nuclear war, so to speak, with both hands tied behind their backs. On the other hand, the provision of really adequate shelters of the types described, and the necessary regimentation of society needed to implement the program, would require an outlay of funds, at a minimum of $5,000 per person, over three times our annual defense budget. It would also—so many persons fear—change the structure of American society irrevocably into that of the closely regimented armed camp, the very image of the totalitarian society we most dislike. An even more immediate threat might be the likelihood that the very sight of our nation engaged in taking such measures of defense would convince the opponent that we had determined to make the first, the "pre-emptive," strike and would thus bring upon us the very horror we most hoped to escape.

THE DELAYED FALLOUT

The problem of worldwide or "delayed" fallout is distinct from that of local fallout. It consists of two components, corresponding to the levels of the atmosphere which the fallout particles reach and from which they descend. The lower level of the atmosphere, the level of winds, clouds, rain and snow, is known as the troposphere. Above it is the stratosphere. The boundary between these is called the tropopause (See Figure 11.1). This boundary is higher in the tropics, where it is found rather constantly at about 55,000 feet, than toward the poles, where it varies between 25,000 and 45,000 feet depending upon the time of day, season, and latitude.

The radioactive particles that enter the troposphere circulate about the earth from west to east, carried by the prevailing winds, until brought down through adherence to raindrops, flakes of snow, or other forms of precipitation. The fallout pattern is often very spotty as a consequence of the irregularities of precipitation. Little of the material produced crosses the equator; it tends to concentrate in the mid-latitudes between

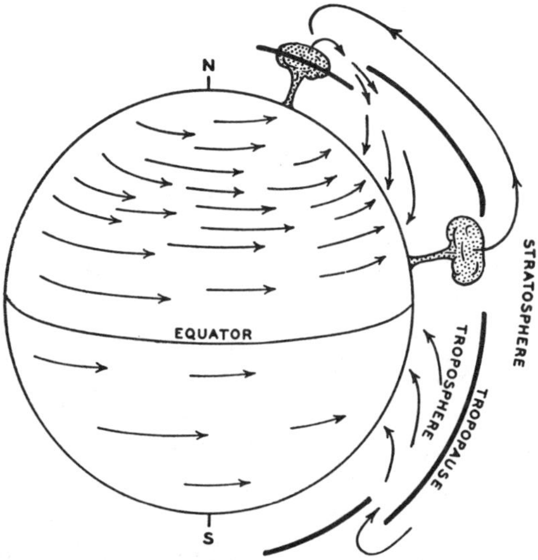

FIGURE 11.1. A diagram to illustrate prevailing winds and air flow between stratosphere and troposphere affecting the distribution of fallout from megaton tests of nuclear weapons in the atmosphere.

30° and 60° N if the detonation has occurred in the Northern Hemisphere. This zone happens to fall, of course, mainly where human populations are dense.

There is a marked break, or gap, in the tropopause in temperate latitudes, and this break moves north or south with the seasons. Through it passes most of the transfer of air from the stratosphere to the troposphere. As a consequence, radioactive material injected into the stratosphere by a nuclear explosion will come down sooner if it is injected in the far north, since it has but a little way to go to the gap in the tropopause, than when injected near the equator, since then it must circulate gradually to the polar regions before it can reverse direction and reach the gap. It follows that the average residence time of fallout material in the stratosphere is very different in the two cases, being about six months in the case of polar injection and about two and one-half years in the case of injection near the equator. The exposure of populations living in the Northern Hemisphere, for any given initial amount of fission products, is of course must greater in the former case. Only a moderate amount of the fission products that reach the stratosphere crosses the equator, not more than 25 percent of the total.

Tropospheric and stratospheric fallout together represent the worldwide fallout, in contradistinction to the local fallout that occurs in the immediate vicinity of an explosion. In the case of a nuclear war such as we are considering, it is the amount of this delayed fallout that will affect the populations of nontarget nations. Had it not been for the monitoring of the fallout from weapons tests conducted before the ban on atmospheric weapons tests, we would have no way of estimating this hazard; but as it is, from the exposure to fallout from the 92 MT of fission products produced by the weapons tests in the period 1952–58, we can extrapolate very simply to the exposures to be expected by neutral nations in the event of a 20,000 MT war involving the United States, the USSR, and probably Western Europe. The population of the United States lives in the latitudes between 30° and 60° N, and because many of the tests were held in Nevada and yielded tropospheric fallout on the United States in particular, the exposure of the American population to fallout is about maximum in the world at the present time. It is stated in *The Effects of Nuclear Weapons*[8] to amount to a whole-body dose of 98 millirems, approximately one-tenth of a rad. This is one-seventieth of the amount of natural radiation received during seventy years, on the average, by members of our population living at sea level; or about 1 percent of the natural radiation received in seventy years at high altitudes such as that of Denver. If half of the 20,000 MT results from fission, the delayed fallout to be expected would be close to 100 times as great as that from the weapons tests during the 1952–58 period. In short, it would amount to about 10 roentgens. Of course, there is some error of measurement in the amount of fallout to which the American population has been exposed to date. The radiation dose might be greater, perhaps even 0.2 rad. Against any such increase of the estimate, however, we must offset several things. First, the nontarget nations may well lie south of latitude 30° N and consequently would receive a lesser

[8]*The Effects of Nuclear Weapons*, S. Glasstone, ed., rev. ed., (Washington, D.C.: U.S. Govt. Printing Office, 1962). Much of the preceding information is drawn from this manual.

amount of fallout. Second, since much of the tropospheric fallout will descend on the countries where the explosions occur, or to the east of them, the larger part of the tropospheric fallout from a nuclear war such as that postulated would spare the Philippines, southern China, Thailand and Burma, India, and Mexico, Central America, Colombia, and Venezuela. And south of the equator less than half as much fallout would be expected.

What, then, would a general dose of 10 rads north of the equator or a dose of 5 rads south of the equator do to the local populations? As we shall see, some radiation damage is to be expected. But many persons in the past have received larger doses of high-energy radiation in a hospital or physician's office, from fluoroscopy or half a dozen radiographs, and no one is *visibly* the worse for it. (With modern equipment and the techniques commonly employed in hospitals and clinics in the United States, doses larger than 5 rads are now rarely delivered in a single fluoroscopy or in half a dozen radiographs. Even so, that is over twice the average annual dose of radiation a person in the United States receives from all background sources.) One may, I think, conclude that even a much larger nuclear war than the one I have taken as a basis of analysis would not wipe out all human life or even destroy all civilization, even in the Northern Hemisphere.

The specific biological effects of fallout are to be gauged in terms of the following ranges of whole-body dose: doses over 1,000 r, death virtually certain; doses around 450 r, death of 50 percent, recovery of 50 percent of persons so exposed (LD_{50}); 200 to 300 r, severe radiation sickness; 100 to 200 r, mild radiation sickness; below 100 r, no evident immediate effects.

The effects of the radiation are plainly most severe upon tissues composed of proliferating cells. When cells are dividing, the hereditary material of the nucleus of each cell condenses. The chromosomes make their appearance, first as long slender threads, each one double as a result of previous replication. The chromosomes thicken, coil up, and become covered with some insulating material. The duplicate members of each original chromosome separate and pass to opposite ends of a spindle, so that they find their way, one of each kind, into the daughter cells. It is in the early stages of this mitosis, as it is called, that the chromosomes are specially liable to fracture and damage by radiation. It has now been clearly demonstrated that most killing of cells by radiation, particularly at doses under 1,000 r, is through damage to the chromosomes, and not to any other parts of the cells.[9] These facts explain why it is *dividing* cells that are killed, and consequently why the primary injuries done by radiation are to the skin, the lining of the digestive tract, the bone marrow, the kidneys, and the reproductive cells. All of these tissues or organs consist of more or less rapidly dividing cells that must frequently replace cells that die. On the other hand, nervous tissue, which is almost totally nondividing after fetal life, is far more resistant to high-energy radiation.

SOMATIC EFFECTS OF ATOMIC RADIATIONS

The effects of radiation of the human body are often divided into somatic effects and genetic effects. These are not really different in the ultimate character of what the radiation does to the components of the cells. The difference is simply that in the case of damage done to reproductive cells, housed in the ovaries or testes (the gonads), any damage done to the chromosomes and genes may be passed on to some future descendant; whereas in the case of damage done in the somatic cells that constitute all the rest of the body, the damage done is not passed on to another generation, but is eliminated with the death of the cells or person receiving the radiation.

Severe radiation sickness begins with nausea and vomiting. Diarrhea and loss of appetite follow. These are symptoms of the killing of innumerable cells lining the digestive tract. Later come hemorrhages, from a loss of blood platelets needed for blood clotting, and a startling drop in the number of white blood cells. Along with these effects there is, naturally, a lowered resistance of the body to bacterial infections and a depression in the formation of antibodies, since the bone marrow and lymph-

[9]T.T. Puck, "Action of radiation on mammalian cells. III. Relationship between reproductive death and induction of chromosome anomalies by X-irradiation of euploid human cells *in vitro*," *Proc. Natl. Acad. Sci. U.S.* 44(1958):772–80.

oid tissues are being depleted. The hair falls out. Unless measures are taken, such as the complete replacement of the bone marrow by implantation from a normal donor, death in severe cases follows after steplike rises in temperature. In the severest cases, death is a consequence of direct degeneration of the tissues after too many of the cells have been killed. In less severe cases, death often comes from infections against which the body has little or no remaining resistance.

Mild radiation sickness is similar except in severity. After the first symptoms develop, they may pass away, and a period of days or even weeks may elapse before there is a recurrence following the first appearance of the blood changes. Antibiotics, to resist the infections that may develop, are effective in these milder cases.

There are several kinds of long-term, or delayed, somatic effects. One of these is the induction of malignancies which may appear after a latent period of a few to as many as thirty-five years. In addition to cancers of lung, stomach, breast, skin, ovary, and other exposed organs, leukemia, which may be described as a neoplasm of the white blood cells, may be produced. Thyroid carcinoma may result from iodine-131; bone cancer may follow the concentration of strontium-90 in the bones. All these sorts of malignancies occur in the population ordinarily, so that what is observed is never more than a statistical increase, often very small and difficult to establish as significant. Yet, even though the difference is small, it has been established as statistically significant in some cases. For example, thyroid tumors are increased in frequency after doses to the thyroid gland of as little as 6.5 rads.[10] Particularly in children, the thyroid is probably one of the most vulnerable organs in respect to carcinogenesis.

Cataracts, resulting in blindness, may arise from opacity of the lens of the eye and are induced either by neutrons or gamma rays in the direct radiation. The flux of such radiation in delayed fallout is probably much too low to produce cataracts. One would judge from the occurrence in the survivors of the Hiroshima and Nagasaki explosions that cataracts follow only

[10]C. Silverman and D.A. Hoffman, "Thyroid tumor risk from radiation during childhood," *Preventive Med.* 4(1975):100–05.

from a sufficiently great exposure (300 rads or more) to the initial nuclear radiation. X-rays, of course, will act similarly, and the exposure of the eyes to large doses must be carefully avoided.

A less obvious somatic effect of radiation is shortening of the normal life span. Children exposed to high-energy radiation may exhibit a slower rate of development and delay in attaining maturity, even when no obvious defects are apparent. But more startling is the fact that irradiated children or adults never live as long as comparable groups of unirradiated persons. This is best shown in the results of some of the experiments done with mice and rats. If animals are irradiated with, say, a 300 rad whole-body dose and are then permitted to recover, or if they are given even a dose insufficient to cause obvious radiation sickness, and if they are then later subjected to a killing dose of high-energy radiation and compared with a group never previously irradiated, it is characteristically found that it takes a considerably smaller dose to kill all the previously irradiated animals (or to kill half, the LD_{50}) than it requires to kill the previously unirradiated ones. One might say that the reserves which the body possesses as a defense against the effects of radiation and upon which recuperation depends have been depleted. Moreover, this depletion is strictly proportional to the amount of radiation previously received; that is to say, the shortening of life is directly proportional to the total accumulated dose.

The most careful and extensive experiments of this kind, to my knowledge, were performed by Patricia Lindop and J. Rotblat in England.[11] Using mice all of a particular age and belonging to a homogeneous genetic strain, they compared groups unirradiated with others irradiated with doses of X-rays ranging from 50 to 780 roentgens (rads). The mice were followed through life until all had died, and each one was then autopsied to determine the cause of its death. The shortening of life produced by the radiation was found to be directly and linearly proportional to the size of the administered dose. There was not the slightest indication that the smallest dose administered

[11]P.T. Lindop and J. Rotblat, "Shortening of life and causes of death in mice exposed to a single whole body dose of radiation," *Nature* 189(1961): 645–48.

produced less than the expected amount of effect. In other words, there was no indication of any threshold below which the radiation fails to shorten life, even though the 50 rad dose is well below the level that produces any visible symptoms. Moreover—a point to be emphasized—the deaths which occurred were from the *usual* causes characteristic of this strain of mice, and not from any unusual causes attributable to the radiation. It is as if the radiation had merely advanced the time of death from the usual causes, quite as if it had removed a segment of the life span at the time when the irradiation was administered and left the irradiated animals biologically older than their chronological age would indicate!

The life-shortening found in the mouse study was 5.66 weeks per 100 rads, or half a day per rad. The LD_{50} for these mice (700 rads) would diminish the life span by 38 percent. Since the LD_{50} for human beings is thought to be closer to 450 rads than to 700 rads, we probably experience a greater life-shortening effect for a given exposure than a mouse does. But then the human life span is also longer—in fact, about twenty-six times as long. Though the extrapolation is none too secure, we might multiply the half day per roentgen by the factor expressing the difference in the life spans of the two species, on the assumption that the mouse simply uses up its biological potential at a faster rate than a human. In that case, one roentgen of whole-body irradiation would be expected to reduce the human life span by thirteen days. If we use the LD_{50} values to obtain the proportional sensitivity of mouse and man, the shortening of the human life span by one roentgen would be expected to be even greater, amounting to 3.2 weeks.

It must be pointed out that the life shortening found by Lindop and Rotblat in irradiated mice exceeds that found by other investigators. This is probably because Lindop and Rotblat used mice at the most radiation-sensitive age of life. This procedure would be comparable to studying the effects on humans of irradiation during early childhood. Adults would prove to be considerably less affected. Yet rough as the extrapolation is, it provides a yardstick for deducing the life-shortening effects of radiation and fallout in the event of nuclear war. Any survivors in the combatant nations will very likely receive 200 to 300 rads of whole-body radiation, and this amount might shorten their lives as much as eight years if they were children, or rather less if they were adult, even if they escaped radiation sickness altogether. As for the inhabitants of noncombatant nations, the 5 rads to 10 rads they might receive would also be expected to produce some degree of life-shortening, but by so few days that in countries where most people do not attain the full natural life span anyway it would not be noticeable. It must be added, however, that the doses used in these mouse experiments were "acute" doses, that is, were administered rapidly. We do not know at the present time whether the slower administration of the same total dose, as from fallout, would produce so severe an effect. There is some reason to suppose that it might not, for a low dose rate yields a lower genetic effect, as will be seen.

Finally, among the somatic effects of high-energy radiations, we must consider induced sterility, for this is a common consequence when sizeable doses of radiation are administered to the gonads. In the case of the mouse, the male becomes sterile in a few days, owing to the killing of the mature and near-mature germ cells; fertility gradually returns, as the more resistant spermatogonia develop and mature into spermatozoa. In human males, a dose as low as 8 rads will kill at least some of the immature germ cells in the testis.[12]

In the case of the female mouse, the opposite pattern occurs. She remains fertile for a time, since the relatively mature oocytes are fairly resistant to high doses of radiation. Then she becomes totally and permanently sterile. Even 50 rads will produce this permanent sterility if administered at high dose rate, or 80 rads administered at low dose rate. If women in this respect respond to radiation like female mice, the most calamitous effect of nuclear war would be the prevention of further reproduction, especially since the younger females in the population would experience total sterility throughout life. Fortunately, the mouse appears to be an exception among mammals, even among rodents, in this respect. Female guinea pigs and hamsters, as well as monkeys, are much more resistant than female mice to the sterilizing effects of radiation. It seems that

[12]M.J. Rowley, D.R. Leach, G.A. Warner, and C.G. Heller, "Effect of graded doses of ionizing radiation on the human testis," *Radiation Research* 59(1974):665–78.

in the female mice the oocytes in the ovary go into a prolonged arrest in development at a more sensitive stage in their maturation process than is the case in the female guinea pig. This difference probably holds true for the other species mentioned, and probably also for the female dog and human female, for which there is evidence that sterility is produced only by much higher doses (several hundred rads). (*Warning*! The very fact that the female's primary oocyte is more resistant to killing by radiation may mean that more of the mutations and chromosome breakages that are induced in them will be transmitted to the offspring.)

If so, it is very unlikely that the women and female children of noncombatant countries, even in the Northern Hemisphere, would exhibit any detectable loss of fertility in the aftermath of a nuclear war of the dimensions I have assumed. In the combatant countries themselves, however, if surviving women have received an accumulated dose of 200 rads or more, even though most of it might be in the post-shelter period at relatively low dose rates, a major effect might derive from the sterilizing effects of radiation. The males would tend to be sterile during the immediate period after exposure; and by the time they were recovering fertility, the adult females would lose their fertility. The younger females might never become fertile. We need much more information about these matters than now exists. Is the reduction of fertility, like the shortening of the life span, proportional to the total dose? In experiments with the fruitfly Drosophila, I have found that even a dose of 5 rads produces a decline of fertility of about 1 percent. This would be of no consequence to a species like the human one, which has a far greater potential fecundity than is needed to keep the birth rate at or above the level of the death rate. In the aftermath of nuclear war, however, piled on innumerable other difficulties and woes, the diminished reproduction of the population is not to be ignored.

THE GENETIC EFFECTS OF ATOMIC RADIATIONS

Three main types of genetic effects produced by ionizing radiation may be distinguished: chromosome breaks, chromosome nondisjunction during cell division, leading to extra or to missing chromosomes; and transmissible gene mutations unassociated with obvious chromosome damage. In previous discussions of the biological effects of radiation, even in the updated 1960 Report of the Genetics Committee of the National Academy of Sciences[13] on this subject, nothing was said about the second of these categories, for it was only in 1959 that evidence was discovered that nondisjunction produces abnormal chromosome numbers in human beings.

Chromosomes each possess one, and normally only one, structure that during cell division attaches it to the spindle and serves as a locomotor organ ensuring that it will move to one pole of the spindle or the other, and thus become properly included in the nucleus of one daughter cell or the other. When a chromosome reproduces itself, this structure (the centromere or kinetochore) also splits or reproduces, so that each daughter chromosome has a centromere at the same location in the string of genes composing the chromosome. When a chromosome or a chromatid (i.e., a daughter chromosome not yet separated from its sister-chromatid) is fractured by radiation, the piece thus separated from connection with its centromere forms a fragment that cannot move appropriately as the normal chromosomes do. The fragment may drift out into the cytoplasm, in which case it is attacked by enzymes that destroy it, and all its genes will then be lost from the daughter cell that otherwise would have received them. The loss of genes may be fatal to the cell, or it may not. In the latter event, however, in human reproduction children that inherit the chromosome from which some of its genes have been deleted nearly always have a neurological defect.[14]

The second thing that may happen, in case more than one chromosome or chromatid is fractured, is that two broken ends may adhere, for they behave as if they were sticky. In that case, if both fragments adhere they will fail to possess a centromere and be lost, as before, while the adhesion of two pieces each of

[13] *The Biological Effects of Atomic Radiation, Summary Reports,* (Washington, D.C.: National Academy of Sciences—National Research Council, 1960).

[14] J. Kunze, "Neurological disorders in patients with chromosome anomalies," *Neuropediatrics* 1(3, 1980):203–49.

which possesses a centromere will produce a chromosome bridge when the two centromeres go in opposite directions. The tug of war that results commonly ends either in the loss of the abnormal chromosome in the cytoplasm of the cell or in rupture of the bridge, which is followed by further losses of bits of chromosome or by further adhesions. Only when broken ends adhere in such a way that every piece of chromosome possesses one, and only one, centromere will a transmissible and heritable effect occur. At low and intermediate doses of radiation there are rarely two independent breaks in the same nucleus. Consequently chromosome breakage at low and intermediate doses, below 1,000 rads, results chiefly in losses of pieces of chromosome, in deficiencies of usually many genes, and in death of the affected cells.

Single chromosome breaks have been shown repeatedly (for example, in experiments done by J. G. Carlson,[15] M.A. Bender,[16] and N. P. Dubinin and collaborators[17]) to increase linearly with the administered dose of high-energy radiation. These experiments have been carried out with human cells of a variety of types, growing in artificial cultures outside the body, as well as in white blood cells irradiated just as they were drawn from the body. Especially clear results were obtained in experiments done in my own laboratory by J.G. Brewen while a graduate student. Brewen[18] used the corneal epithelium of the Chinese hamster, irradiated with X-rays in its normal position in the eye. These experiments, like those of Bender and of Dubinin and his coworkers, indicate that the frequency of chromosome breakage is linearly proportional to the dose in roentgens, at relatively low doses. Brewen's data cover the range from 10 rads to 150 rads very adequately. They showed an

[15] J.G. Carlson, "An analysis of X-ray induced single breaks in neuroblast chromosomes of the grasshopper (*Chortophaga viridifasciata*)," *Proc. Natl. Acad. Sci. U.S.* 27(1941):42–47.

[16] M.A. Bender, "X-ray induced chromosome aberrations in mammalian cells *in vivo* and *in vitro*," in *Immediate and Low Level Effects of Ionizing Radiation, Intern. J. Radiation Biol.*, Suppl. (1959):103–18.

[17] N.P. Dubinin, Y.Y. Kerkis, and L.I. Lebedeva, "Experimental analysis of the effect of radiation on cellular nuclei in the culture of embryonal human tissue," *Soviet Physics Doklady* 6(1961):538–40.

[18] J.G. Brewen, "Dependence of frequency of X-ray-induced chromosome aberrations on dose rate in the Chinese hamster," *Proc. Natl. Acad. Sci. U.S.* 50(1963):322–29.

average frequency of breakage of .0035 per cell per roentgen (rad). Bender's results on the white blood cells yield a frequency of .0024 per cell per rad. For tissues comparable to these, a dose of 300 to 400 rads will produce one chromosome break, usually cell-lethal, in every cell. It is easy to see why doses in this range are so drastic in effect. The miracle is that any individual so exposed can survive!

Because the production of a chromosome break is usually lethal to the cell or to one of the two daughter cells, such effects are not likely to be passed on to the next generation. The principal results will be to add to the sterility burden produced by other effects of the radiation. Clearly, since reproductive cells are cells that must divide, except for those already mature, which are likewise quite sensitive to radiation, a dose of 300 to 400 rads would seem likely to produce a grave loss of fertility. Even a dose of 10 rads might be expected to reduce fertility proportionately, perhaps by 2 to 4 percent. Experiments done with the fruitfly Drosophila in my own laboratory confirm this prediction. The fertility of unirradiated males and females was compared with that of males and females exposed to a dose of 5 rads each. In a total of 1,360,948 progeny, there was a 1 percent reduction in the number of offspring produced by the irradiated parents.

The second type of genetic damage which may be produced by radiation is a consequence of errors that result in an extra chromosome of some sort, or one chromosome too few, being present in the fertilized egg or in one of the cells arising from it early in the course of development. In 1959 it was discovered, almost simultaneously in France and in Great Britain, that most cases of Down's syndrome, characterized by severe mental defect, are attributable to the presence of an extra chromosome, designated number 21 from its relative rank in size among the twenty-three chromosomes of egg or sperm. Shortly afterwards, two forms of sexual maldevelopment accompanied by sterility were discovered to arise, the one from the presence of an extra sex chromosome, the other from the lack of a sex chromosome normally present. Ever since 1913 abnormalities of this same kind had been known in fruitflies, and later they were found (in 1922) to be increased in frequency after the parent flies were exposed to X-rays. Similar abnormalities have more recently been found in mice.

These errors of chromosome number, which may be classed as a sort of mutation, arise in two general ways. The commonest is probably through what is called nondisjunction, that is, a failure of the maternal and paternal chromosomes of the same kind, after their close pairing during the maturation of the germ cells, to separate from each other and to go singly into the reproductive cells. The result of the error would be formation of one reproductive cell with an extra chromosome and another cell with one chromosome too few. The other way in which such errors arise is through the loss of a chromosome from the nucleus of the fertilized egg. Studies at the Oak Ridge National Laboratory by Liane B. Russell and C.L. Saylors show that when mouse spermatozoa are irradiated, either before or after the transfer to the female during mating, loss of a chromosome is not uncommon. In fact, the frequency of loss of one of the paternal chromosomes is considerably greater when the irradiation hits the paternal nucleus after the sperm has already penetrated the egg during the fertilization process than when the sperm is X-rayed in the male parent prior to mating. In the former case, 5.2 percent of losses of the sex chromosome occurred after a dose of 100 rads, in the latter case only 0.2 percent.[19]

Losses of chromosomes other than the sex chromosomes are fatal during early development in the mouse, and in human beings losses of most chromosomes except the sex chromosomes and the smaller chromosomes (numbers 21 and 22) are likewise fatal. Many particular kinds of loss have now been identified and are known to cause multiple congenital defects that result in neonatal death. The amount of prenatal death from such causes is difficult to determine, but it is certainly great. The proportion of embryos and fetuses that perish from miscarriage or natural abortion is variously estimated as ranging up to 50 percent or more. Most of those which have been recovered for examination reveal chromosome abnormalities such as result from nondisjunction or from chromosome breakage. This proportion seems far too high to be caused exclusively by the natural radiation in the human environment. Like gene mutations, most of them must arise from other causes. But it is nevertheless still true that any large increase in the dose of

[19]Liane B. Russell and C.L. Saylors, "Induction of paternal sex chromosome losses by irradiation of mouse spermatozoa," *Genetics* 47 (1962):7–10.

radiation received by the parents will be reflected in an increase in the proportionate numbers of prenatal deaths caused by chromosome abnormalities.

In any event, all previous estimates of the frequency of detrimental mutations produced by ionizing radiations must now be increased by some, still indefinite, amount to allow for this novel and unsuspected type of human damage. At the present time one cannot say how many aberrant persons, as well as fetal and neonatal deaths, have been produced by radiation, apart from other causes. It may indeed be a small proportion of the total. Yet there is no reason to doubt that radiation will cause such defects, particularly if administered to the female parent or the just-fertilized egg. The continuous radiation from the surroundings after a large-scale nuclear attack might produce a notable increase of this sort of mutational defect, even though in lands where the radiation from delayed fallout was 10 rads or less no difference from previous conditions in this respect would be apparent.

Transmissible mutations induced by high-energy radiations include both small deficiencies, usually at the lower limit of detection even in the giant salivary gland chromosomes of the fruitfly, and also what the geneticist calls "point mutations," cases where the alteration in the chromosome is below the resolution of current microscopy. Both of these types of mutation are related to the dosage of radiation in the same way. They increase in direct linear proportionality to the dose. That is to say, double the dose, and you obtain twice as many mutations; triple the dose, and you obtain three times as many; and so on. Much effort has gone into the establishment of this relationship for relatively low doses of radiation, since the dosage curve, when extrapolated downward from the experimentally established levels to lower doses, always appeared to hit at its origin (at zero dose) the level of the established spontaneous mutation rate. In other words, there was no evidence of any threshold below which the radiation was ineffective in producing mutations. To the contrary, one would conclude that every increment of radiation, however small, produced its proportional effect upon the mutation rate.

Most early studies of induced mutation were limited to radiation of the male and usually to treatment of mature spermatozoa. Later it became clear that such results do not give a sufficient picture of the effects of radiation upon a population, because they neglect the different responses of the mature females and particularly of the immature germ cells (which might be carried either in immature individuals or in mature individuals). The wartime studies of the Manhattan Project were, in fact, carried out principally with male fruitflies. Teams of geneticists and technical assistants succeeded in showing, first, that 50 roentgens, and later, 25 roentgens, produce exactly the frequencies of mutations expected in direct, linear proportion to those obtained at higher doses.[20] Notwithstanding, as concern grew about the accumulated effects of fallout or of the doses given for medical and dental diagnosis and treatment, there were doubters who wondered if 10 R, or 5 R, or 3 R, would still produce the expected number of mutations, or even any at all. For this reason, my colleagues and I carried out, over a period of three years, what we sometimes called our "megafly experiment," since we examined considerably more than a million files for mutations induced by a dose of 5 R given to each of the parents. The experiment was repeated fifty times, each time with exactly matched irradiated and unirradiated cultures coded to prevent any possible bias on the part of the scorers had they known which series had been irradiated. A dominant type of transmissible mutation occurring at more than fifty loci in the chromosomes was the basis of the test. These minute-bristle mutations are either point mutations or small deficiencies, and earlier tests had shown that, for a given dose of X-rays, they arise with equal frequency in spermatozoa and in mature oocytes. The results showed definitely that even a 5 R dose produces mutations at the proportion expected on the basis of a linear relation to dose; that is, 5 percent as many as from a dose of 100 R, or one-half of one percent as many as from a dose of 1,000 R.[21]

[20] W.P. Spencer and C. Stern, "Experiments to test the validity of the linear r-dose/mutation frequency relation in *Drosophila* at low dosage," *Genetics* 33(1948):43–74. E. Caspari and C. Stern, "The influence of chronic irradiation with gamma-rays at low dosages on the mutation rate of *Drosophila melanogaster*," *Genetics* 33(1948):75–95. Delta E. Uphoff and C. Stern, "The genetic effects of low intensity irradiation," *Science* 109(1949):609–10.

[21] B. Glass and Rebecca K. Ritterhoff, "Mutagenic effect of a 5-r dose of X-rays," *Science* 133(1961):1366. (A preliminary abstract).

Together with the evidence from the production of chromosome breaks by low doses in individual cells and tissues, this evidence reinforces what was put in carefully chosen words by the Genetics Committee of the National Academy of Sciences Committee on the Biological Effects of Radiation in 1956: "*Any* radiation is genetically undesirable, since any radiation induces harmful mutations.... The genetic harm is proportional to the total dose."[22] And, in the 1958 Report of the United Nations Scientific Committee on the Effects of Atomic Radiation: "Even the smallest amounts of radiation are liable to cause deleterious genetic, and perhaps also somatic, effects."[23]

The most recent studies by the United Nations Scientific Committee on the Effects of Atomic Radiation and by the Advisory Committee on the Biological Effects of Ionizing Radiation (the BEIR Committee) of the National Academy of Sciences, in 1977 and 1980 respectively, continue to hold to the view that there is no threshold dose in respect to the induction of genetic mutations or chromosome breaks by ionizing radiations. In 1982, Arthur C. Upton wrote a lucid and splendidly illustrated article about the biological effects of low-level ionizing radiation, especially the risks of various kinds of cancer. He summed up present scientific and medical opinion in these words:

> According to the view that no amount of radiation is entirely without effect, some risk must be presumed to be associated with any activity involving low-level radiation, no matter how small the dose may be. Some such risks, compared with the other hazards of daily life, seem negligible. Still, no risk can be regarded as acceptable if it is readily avoidable or if it is not accompanied by a commensurate benefit.[24]

The most unexpected and most discussed development during recent years in the study of radiation-induced mutations

[22]*The Biological Effects of Atomic Radiation, Summary Reports* (Washington, D.C.: National Academy of Sciences—National Research Council, 1956), p. 23.

[23]*Report of the United Nations Committee on the Effects of Atomic Radiation* (New York: United Nations, 1958), p. 41.

[24]Arthur C. Upton, "The biological effects of low-level ionizing radiation." *Scientific American* 246 (Feb. 1982): 41–49.

has emerged from the mouse experiments conducted by W.L. Russell and his colleagues at the Oak Ridge National Laboratory.[25] These experiments were done by irradiating mice lacking any known mutations and then mating them to a strain which carries seven different recessive point-mutations. The offspring of such a cross would be heterozygous, that is, would carry at each of the seven loci one normal and one recessive mutant allele. They would therefore be nonmutant in appearance unless a mutation—either a point-mutation or a deficiency—had been induced by the radiation at one of the seven sites. The earlier experiments were performed by treating the male with X-rays, waiting until it had fully recovered its fertility, and then mating it to the tester type of female. The results, over a range of doses from 300 rads to 1,000 rads showed the expected increase in frequency of mutations to be linearly proportional to the increase in dose. The later experiments compared the results just described with those obtained when a gamma-ray emitter such as cobalt-60 or cesium-137 was used instead of X-rays, and the rate at which the dose was delivered was greatly reduced. The earlier dose rate was 90 rads per minute and the new one was 0.009 rad per minute, that is, one ten-thousandth of the higher rate. The total doses administered were of course adjusted to be the same. To everyone's surprise, since previous experiments in producing mutations at different dose rates had shown no differences between the effects of high dose rates and low dose rates whenever the total dose was the same, there now appeared to be a considerable difference. At the low dose rate only about one-fourth as many mutations were produced as at the high dose rate, for each and every total dose.

The reason why the effect had never been found before is now reasonably clear. The earlier experiments, which were performed largely with fruitflies, had involved only the treatment of mature spermatozoa, whereas by Russell's method the treated cells which functioned in fertilization were immature germ cells (spermatogonia) at the time of their treatment. Spermatozoa, although very sensitive to ionizing radiation, are

[25] W.L. Russell, Liane B. Russell, and E.M. Kelly, "Dependence of mutation rate on radiation intensity," *Intern. J. Radiation Biol.*, Suppl. (1960): 311–20.

nondividing cells; but spermatogonia are dividing cells, and the conditions prevailing in them evidently allow some sort of repair or restoration to occur in the case of many mutations that, when they occur in spermatozoa, are unable to undergo repair.

Later studies by Russell's group indicated that the effect of the dose rate also exists in the maturation of the female reproductive cells, or oocytes, and may in fact be greater in magnitude in them than in the male. For both sexes together, about six times as many mutations are induced at a high dose rate as at a low dose rate. The implications of these findings for the interpretation of genetic damage from fallout are obvious. In the event of a nuclear war, much of the exposure of the survivors in a target country will occur at fairly high dose rates from the local fallout within the first two days after the nuclear explosions. Perhaps half of the exposure, however, would come from delayed fallout; and in this case the dose rate will be relatively low. On the other hand, exposed populations in noncombatant countries will be exposed entirely to delayed fallout, that is, to radiation at a low dose rate. If the human reproductive cells respond to radiation as do those of a mouse, mutations will be considerably less frequent under these circumstances.

Russell has also tested several intermediate levels of dose rate. In the experiments in which males are treated, a dose rate of 9 rads per minute is intermediate in its genetic effect, while a dose of 0.8 rad per minute is already at the minimal effectiveness of a dose rate a hundred times as slow. When females are treated, the situation turns out somewhat differently. The minimum level is not reached until a dose rate below 0.8 rad per minute is utilized.[26]

Several important conclusions may already be drawn, although confirmation of these results, particularly for other species of animals, is still needed. One point to emphasize is that the male and the female reproductive cells cannot be expected to behave toward radiation exactly alike. The situation is complex. Secondly, one should emphasize that in terms of human exposure, 0.8 rad per minute is not what would be

[26]W.L. Russell, "Effect of radiation dose rate on mutation in mice," *J. Cell. Comp. Physiol.* 58 (Suppl.) (1961):183–87.

considered a very low dose rate, since at that rate it would require only twelve to thirteen minutes to reach the level of exposure that is currently the accepted maximum permissible limit for gonadal exposure of the population to man-made radiation. Thirdly, there is no ground, on the basis of Russell's data, for concluding either that the relation of mutation frequency to dose is not one of linear proportionality or for concluding that there is any threshold at very low doses. In both of these important respects, the data show precisely the contrary. For the low dose rates, the frequency of mutations still increases linearly with the dose. A total dose of only 86 rads at the low dose rate produced exactly the frequency of mutations expected proportionately on the basis of the frequencies of mutations produced at doses of 300 rads, 516 rads, and 861 rads, also administered at the low dose rate. There is no indication of a threshold.

Several recent discoveries tend to offset the lowering of the estimated genetic damage because of the dose-rate findings. Especially important is the induction of mutations by carbon-14, which is produced in the fusion process, has a long half-life of over 5,000 years and is readily incorporated into the genetic material of the very genes and chromosomes themselves. Estimates by geneticists were given in the Report of the Federal Radiation Council (1962) entitled *Health Implications of Fallout from Nuclear Weapons Testing through 1961*.[27] Whereas in the first generation of children of persons exposed, the number of gross physical or mental defects arising from mutations produced by carbon-14 was estimated to be only one-tenth of the number produced by the total delayed fallout, the total number through all future generations was estimated to be twice as great as the number arising from the delayed fallout. Local fallout from nearby explosions is of course not meant here, since the estimates were based on the effects of fallout produced by weapons testing to date. Since the survivors in an attacked country would probably receive a large part of their exposure from delayed fallout, the effects of carbon-14 will by no means be negligible.

[27]*Health Implications of Fallout from Nuclear Weapons Testing through 1961*, Report No. 3 (Washington, D.C.: Federal Radiation Council, 1962).

What would a dose of 200 roentgens (rads) received by every survivor in an attacked country produce, in genetic terms? From James V. Neel's estimates presented in the 1959 *Hearings* we can get some idea.[28] If we assume a constant size of population, 40 million survivors might produce in the succeeding thirty generations some 1,200 million descendants. Among these, to go by Neel's minimum estimates, there would be about 10 million deaths occurring at or before birth from genetic defects, about half a million living defectives, and about 25 million persons with impaired vigor or fertility. In Western Europe and the USSR there might be in the aggregate four times as many. In the descendants of the 2 billion persons living in noncombatant parts of the Northern Hemisphere, the estimated 10 rads they might receive on the average would yield 25 million deaths at or before birth, one and a quarter million living defective persons, and 62.5 million individuals with impaired vigor or fertility. Because of the enormous difference in the sizes of the surviving populations, the small dose from delayed fallout would in the long run produce almost as many genetic disabilities and deaths in the noncombatant peoples of the world as in the survivors of the nuclear combat.

The estimate may be made in another way. The number of genetically defective persons born alive probably amounts to 4 percent of all births. This estimate, somewhat higher than the one used in the National Academy of Sciences Genetics Committee's Report of 1956, is based on the tabulation of specific hereditary traits and their estimated incidences given in the Report of the United Nations Scientific Committee on the Effects of Atomic Radiation, Annex H, in 1958. By genetic defects we mean, of course, not gross monsters in the majority of cases, but the more common kind of biochemical defects, the so-called "inborn errors of metabolism," which are already present in the population and some of which are familiar to any of us. They would include, for example, defects of the gastrointestinal tract, such as intestinal polyps; of the blood,

[28] J.V. Neel, in *Biological and Environmental Effects of Nuclear War*, Hearings before the Special Subcommittee on Radiation of the Joint Committee on Atomic Energy, Congress of the United States, 86th Cong. June 22–26, 1959. (Washington, D.C.: U.S. Govt. Printing Office, 1959), pp. 610–14.

such as hemophilia; of immunity mechanisms, such as agammaglobulinemia; of kidney function, such as cystinuria; nervous and mental disorders, such as epilepsy and schizophrenia; muscular and skeletal defects, such as muscular dystrophy; endocrine disorders, such as pituitary dwarfness or diabetes; reproductive effects, producing sterility; and a long list of others, from albinism and achondroplastic dwarfness through deafness, gout, and myopia, to xeroderma pigmentosum and xanthoma tuberosum.

By a recognized genetic principle, a doubling of the mutation rate will in time lead to a doubling of the amount of evident genetic defect in the population. The question, then, is how much high-energy radiation will double the previous (spontaneous) mutation rate that is responsible for the genetic defects currently present in the population. For a variety of organisms (including some plants, the fruitfly Drosophila, and the mouse), doubling doses for transmissible mutations of the point mutation or small deficiency types have been determined. There is astonishing uniformity in the range of this so-called "doubling dose." It generally falls within the range of 30 to 80 rads, with 40 to 60 rads more common. In my own megafly experiment, described previously, the doubling dose turned out to be 60 rads. For chromosome breaks it is of course much less, but that is a matter of nontransmissible types of mutation. One cannot assert positively that a doubling dose of 40 to 60 rads applies to human beings, but there is no reason to suppose that it does not. One would therefore conclude that a 10 rad dose to an entire population might produce one-fourth to one-sixth of the number of harmful mutations that occur spontaneously in every generation. A dose of 200 rads, on the other hand, would produce three to five times as many as normally arise spontaneously.

The number of genetically defective individuals would consequently rise in the populations exposed to delayed fallout to approach a new level of 5 percent or somewhat less. In the surviving populations of the target countries, on the other hand, the expected final level would rise to 12 to 20 percent of births. True, this would not occur until many generations had passed, and if the exposure were limited to a single generation the level would gradually drop back again to the original level.

Could a population withstand the social effects of such a high level of genetic defect? It probably could; although in the harsh conditions of life that would very likely prevail in the devastated parts of the world after a nuclear war, it is scarcely conceivable that our present tender regard for the preservation of life could continue to be manifested toward the handicapped, deformed, and weakly individuals who would constitute so tremendous a social burden. In the bitter struggle for the necessities of life, what is likely to become of the ethics of modern man? One of the gravest losses which society might experience could, under these conditions, be a reversion to the ethics of the ape man.

A question often asked is the following: "Aren't some mutations beneficial? Can't an increase in the number of mutations be expected to produce some increase in human evolution, some improvement in the overall capacities of mankind?" The question reveals a grave misunderstanding. Some mutations are indeed beneficial, and some that are detrimental in certain environments may be less so, or might even be beneficial, in other circumstances. Evolution is based on the occurrence of such mutations. But it is equally important to understand that the great majority of mutations produce defect and do harm, if for no other reason than because natural selection over countless generations has already been at work selecting and grouping into harmonious genotypes those mutations of which human genes are capable. To the extent that we are well adapted to our present environment, genetic changes are likely to produce a lower fitness. Such effects of mutation as have been analyzed biochemically show that the usual consequence of mutation is to alter the structure of some enzyme or to cease producing it. In either case, the efficiency of some particular step in the body's metabolism is impaired or blocked altogether. Genes control the nature of our enzymes—the enzymes needed in growth, development, and maintenance—and we do not possess very many, if any, superfluous enzymes. Geneticists are therefore not surprised to find, in actual experiments, that the overwhelming majority of detected mutations are damaging. A general increase in the frequency of mutation therefore multiplies injury far more rapidly than benefit. Only selection can turn the tables, and for it to do so requires generations of the elimination of the "unfit," that is, the less viable, the less reproductive individuals. To gain one beneficial gene, a population has to suffer many "genetic deaths," or eliminations of damaging genes. The choice, therefore, is reduced to this: how high a price in present suffering do we want to pay for some possible evolutionary gain? It is likely that, in large populations, beneficial mutations of nearly every possible sort already exist, though perhaps not in ideal combinations. No one, for the sake of one advantageous trait, would want to be handicapped with twenty defects. We can do considerably better by keeping the mutation rate low, even if long-range improvement of the species is thereby postponed.

The damage to a population is nevertheless not simply a matter of the frequency of the detrimental mutant genes. It also depends on the relative amounts of harm they do to individuals and to society. The 1960 Report of the Genetics Committee of the National Academy of Sciences asked provocatively: "How... does one measure quantitatively the relative importance of a still-birth, a feebleminded child, and a death during adolescence?"[29] Or, one might add, how do we evaluate the importance of a death very soon after conception, when the mother is often quite unaware that an abortion has occurred? A tentative beginning of this kind of appraisal has been made by Sewall Wright in an addendum to the report just mentioned. Much thinking and much investigation must go into this kind of combined genetic and social analysis, however, before appropriate measures of any kind can be recommended.

THE AFTERMATH

One cannot examine the consequences of nuclear war in general biological terms and rest satisfied with an estimate of the immediate effects upon the human population. Man lives in a well-balanced environment, and the profound effects of a nuclear war upon the unsheltered and unshielded organisms that make up his ecological community dare not be forgotten.

[29] *The Biological Effects of Atomic Radiation, Summary Reports,* (Washington, D.C.: National Academy of Sciences—National Research Council, 1960), p. 11.

However adequate our shelters for people, there must remain outside them the animals we have domesticated, the plants we cultivate, the forests and the meadows, the soil, the birds, the insects, and the bacteria—to mention only a part of our community.

The outside levels of radiation, in the event of a 10,000 MT attack on the United States, will obviously far exceed lethal limits for birds and mammals. Barring special measures for their protection, survivors will find no horses, cattle, pigs, chicken, or other animal sources of food remaining. Without the birds, the insects will take over. Since adult insects are in general very resistant to high-energy radiation, they will remain in large numbers wherever fire has not completely destroyed life. A cockroach, for example, can tolerate 400,000 rads, and many insects will tolerate 40,000 rads or more. The larval forms are more sensitive and might be killed were it not that many of them live and grow under water or under the bark of trees or in similar sheltered places. And if all the larvae perished but the adults were not rendered completely sterile, the populations would soon be replenished. Sterility, in insects, is not produced by doses of less than several thousand roentgens, which is far more than a mammal or bird can tolerate and live and more than many insects in the continental area are likely to receive as a dose. The insects, unchecked in their ravages, may quickly destroy every green thing left by fire and blast.

The incendiary effects of nuclear weapons are increased by exploding them at high altitudes. According to one estimate, one 10 MT bomb could spread searing heat over 5,000 square miles, which is about the area of the state of Connecticut. The firestorms and lesser fires would hit countryside as well as cities and, depending upon the season of the year and the dryness of the vegetation, might burn off the land over vast stretches. What the fires left, the local fallout would probably kill, since a pine tree can stand only about 500 roentgens (rads), and hardwood trees less than 1,000 rads. Not only would they be destroyed, but unless somewhere seeds were preserved they could not be replanted. The erosion that would follow on the unprotected soil would be frightful to think of. It would probably carry off most of the fallout particles, so that the long-term accumulation of strontium-90 and cesium-137 in the bodies of the survivors might be far less than one would estimate from an initial contamination running up to 100,000 millicuries per square mile in some areas. But the floods and the loss of the topsoil from most of the continent would be a severe price to pay for nature's cleanup of the radioactive contamination.

Finally, one must think of the bacteria. There are so many bacteria that even at high doses of ionizing radiation some escape its killing effects. Unless the fires did a very thorough cleanup job the millions of unburied corpses of human beings and animals would provide a hotbed for the growth of the organisms of death and decay. Can we believe that in the absence of organized medical services, in the wake of physical and psychic stress, radiation injuries, the destruction of sanitary facilities, and malnutrition, disease would not become rampant? Until recently, little serious thought had been given to the prevention of the epidemics that might sweep the land. I have heard it seriously suggested that it would be better, in case of a heavy nuclear attack, if survivors were relatively few and scattered, since in that case there might be a better chance to escape from epidemics, to feed the survivors, and to rebuild a society, than if too many persons were left alive and needing care.

The disorganization of society, the destruction of the laboratories and libraries of science, the obliteration of organized education, the loss of art, music, books, and entertainment must be added to the problems raised by the wreckage of transportation and communication and the loss of natural resources.

The recent projections by Paul Ehrlich, Carl Sagan, and others of a "nuclear winter" that might result from the vast quantities of soot and dust thrown into the atmosphere by massive nuclear explosions, blanketing the earth from most sunlight, radically lowering terrestrial temperatures, and putting an end to nearly all photosynthesis upon which men and animals depend for their oxygen as well as their food, merely heighten with overkill the desolation of the environment. The nuclear winter is by no means a certainty at this juncture in our knowledge, but neither can it be dismissed as simply the product of wild imagination. Were it to come, only the refuge to be sketched in the final section of this essay could possibly save humanity for the several years it would take for the environment to recover.

A RATIONAL CIVIL DEFENSE

(IN RECOLLECTION OF DEAN JONATHAN SWIFT'S "MODEST PROPOSAL...")

What, then, might be done to avert utter collapse, beyond hope of repair, if our country is subjected to a nuclear attack? No system of above-ground evacuation or provision of shelters can possibly be more than a death trap. Well-stocked, appropriate shelters in the more remote rural areas may enable some few families to survive, but we must give up the idea that cities and towns can escape the terrifying effects of blast, firestorm, and heavy local fallout. Virtually all hospitals, clinics, and pharmacies will be destroyed—and, of course, most of the physicians and nurses will themselves be dead or casualties.

First of all, there must be an intensive training of the general population in regard to an elementary understanding of radiation sickness and as to precautions against exposure to fallout. To this date, none of our governmental health agencies has been willing to face this necessity, probably in fear of alarming the population too greatly. Yet isn't that, to a great extent, just what is needed? Complacency, from the highest levels of our government down, is the greatest guarantee that we will meet the utter finality.

More than education is needed, of course, if a bombed population is to survive, at whatever primitive level of existence remains possible. Deep underground, preferably hewn from solid rock, there should be constructed a great system of medical-biological laboratories where vitamins, drugs, and medicines can be stockpiled, antisera prepared, vaccines preserved. Great quantities of stored blood of all types, and artificial blood to supplement the supply, cannot be forgotten. Seeds of all our important crop plants and vegetables, in the rich variety of their cultivated strains, must be preserved as permanently as possible, be grown in sufficient measure to harvest fresh yields of seeds and be restockpiled. The sperm and ova of our domesticated animals must be frozen and preserved, along with very young embryos that can be implanted in foster mothers for delivery and rearing. Of course, that implies that a sufficient number of young and vigorous females of each species must be maintained underground to serve as mothers, even though, in the pressure for space, most males will be dispensed with. Very active research must be carried forward in the underground laboratories in order to speed the day, now not so remote, when animal embryos can be reared without a natural womb, but in a device of glass and plastic and nutritive fluids instead.

The problem of how to save the bird population, which will be needed so desperately to cope with the anticipated ravages of insects in the postnuclear environment, is a difficult one. Likewise, to save our tree species so that replanting after a safe interval can check erosion and preserve the essential soil, is not easy. Probably the seeds of trees and young seedlings can be kept in the underground storehouses, and some may even be reared in artificially illuminated tree nurseries deep below ground surface. To do so will require knowledgeable foresters in those laboratories, as well as a safe, unintermittent supply of electricity for providing light and heat. Inasmuch as aboveground powerplants of all sorts will have been destroyed, and fuel is bulky to store underground, probably the underground laboratories will each need to depend upon its own nuclear energy plant.

In view of the probably enduring sterility and the genetic damage of survivors above ground, the human population too must be prepared for underground reproductive perpetuation. Sperm banks and ovum banks, as well as banks of frozen postfertilization embryos, must be kept for implantation in surrogate mothers in sufficient numbers. As in the case of domestic mammals, male individuals are less necessary for the perpetuation of the species if there is an adequate and varied supply of frozen semen. Our "1984" world in the underground may therefore be manned largely by trained and dedicated women who can serve as the surrogate mothers.

The list of what is needed for such laboratory shelters has scarcely begun. Food stores and running water are obvious. But if civilization is to be saved, more than the perpetuation of people is involved. All the technical and cultural information of the world must be preserved as fully as possible. In this day of computers that are giant in capacity yet grow smaller and smaller in size, much can be stored in them. The underground laboratories will need their own computer systems, as well as complete libraries of all extant books reduced to microform for compact storage. Bulky art may be impossible to save, but literature and reproductions of great art on paper or film can be stored economically.

Finally, of course, since no such imaginary underground city can be perfectly immune to destruction, a country should have two or three storehouses of civilization, operating independently if necessary but interconnected by subterranean cables at great depths. The project sounds fantastic, yet if nuclear war is an imminent threat, just one year's national defense budget—withdrawn from our useless bombs, missiles, and antiquated weaponry—would perhaps suffice to provide for it. So much could be saved even by a halfway agreement for disarmament that the project could be carried through. Much good planning will be necessary to flesh out the bare bones of this solution, but there seems to be no insuperable obstacle. The economy of the country would be vastly strengthened by the project, and, as in the genesis of the Manhattan Project, many jobs would be created.

Long ago a great prophet said: "Without vision the people perish." We rely on our political leaders, but they, most unfortunately, have almost without exception a lack of understanding of the consequences of nuclear war as it would actually be waged between great enemy powers. The medical and biological consequences have received almost no consideration from military, civil defense, or political leaders. There is no preparation worthy of the name to ensure a real survival of America. If, then, the war comes, in our present state of unpreparedness, what is to be expected?

Life will be very primitive for the survivors for a long time to come. If America survived at all, could it be as more than a tenth-rate power? Can one believe that the more fortunate lands would take pity on the country that first produced and used the atom bomb and later engaged in the accelerating arms race? The preservation of the American people and their society, like that of the USSR and Western Europe, demands sincere and continued efforts to attain the general and complete nuclear disarmament of all nations.

Index

Index

About the Author

Bentley Glass, born in 1906 in China, received his undergraduate education at Baylor University and received his Ph.D. from the University of Texas. He taught at Stephens College, Goucher College, the Johns Hopkins University, and the State University of New York at Stony Brook, at the last of which he was academic vice-president for six years. He is a member of the National Academy of Sciences, the American Philosophical Society, and the Czechoslovakian Academy of Sciences. He is the author of more than 300 articles in genetics and general biology and has written or edited twenty-four books. He was the first chairman of the Biological Sciences Curriculum Study and devoted much time to national and international education in science. He is a past president of the American Association for the Advancement of Science, the American Association of University Professors, Phi Beta Kappa, the American Society of Human Genetics, the National Association of Biology Teachers, the American Society of Naturalists, and the American Institute of Biological Sciences. He has lectured nationally for Sigma Xi and for Phi Beta Kappa. Since his retirement from academic life, he has directed a program in the history of genetics at the American Philosophical Society.

About the Founder of This Series

Ruth Nanda Anshen, Ph.D., and Fellow of the Royal Society of Arts of London, founded, plans, and edits several distinguished series, including World Perspectives, Religious Perspectives, Credo Perspectives, Perspectives in Humanism, The Science of Culture Series, The Tree of Life Series, and Convergence. She also writes and lectures on the relationship of knowledge to the nature and meaning of man and to his understanding of and place in the universe. Dr. Anshen's book, *The Reality of the Devil: Evil in Man,* a study in the phenomenology of evil, is published by Harper and Row. Dr. Anshen is a member of the American Philosophical Association, the History of Science Society, the International Philosophical Society, and the Metaphysical Society of America.